T0301822

Integrated Miniaturized Materials—
From Self-Assembly to Device Integration

MATERIALS RESEARCH SOCIETY
SYMPOSIUM PROCEEDINGS VOLUME 1272

Integrated Miniaturized Materials— From Self-Assembly to Device Integration

Symposium held April 5–9, 2010, San Francisco, California, U.S.A.

EDITORS:

Carlos J. Martinez
Purdue University

David Gracias
Johns Hopkins University

Amit Goyal
Oak Ridge National Laboratory

Rajesh R. Naik
Air Force Research Laboratory

Alberto Saiani
University of Manchester

Symposia Organizers:

Sonia Grego
RTI International

Laurent Jeannin
Solvay SA

Alberto Fernandez-Nieves
Georgia Institute of Technology

William Murphy
University of Wisconsin–Madison

Joao Cabral
Imperial College London

Oliver G. Schmidt
IFW Dresden

Qinghuang Lin
IBM T.J. Watson Research Center

Paul W.K. Rothemund
California Institute of Technology

Jeffrey J. Urban
Lawrence Berkeley National Laboratory

Bartosz A. Grzybowski
Northwestern University

James J. Watkins
University of Massachusetts

Rein V. Ulijn
University of Strathclyde

Joel Collier
University of Chicago

Molly M. Stevens
Imperial College London

Athene Donald
University of Cambridge

Phillip B. Messersmith
Northwestern University

Materials Research Society
Warrendale, Pennsylvania

CAMBRIDGE
UNIVERSITY PRESS

Shaftesbury Road, Cambridge CB2 8EA, United Kingdom

One Liberty Plaza, 20th Floor, New York, NY 10006, USA

477 Williamstown Road, Port Melbourne, VIC 3207, Australia

314–321, 3rd Floor, Plot 3, Splendor Forum, Jasola District Centre, New Delhi – 110025, India

103 Penang Road, #05–06/07, Visioncrest Commercial, Singapore 238467

Cambridge University Press is part of Cambridge University Press & Assessment,
a department of the University of Cambridge.

We share the University's mission to contribute to society through the pursuit of
education, learning and research at the highest international levels of excellence.

www.cambridge.org
Information on this title: www.cambridge.org/9781605112497

Materials Research Society
506 Keystone Drive, Warrendale, PA 15086
http://www.mrs.org

© Materials Research Society 2010

This publication is in copyright. Subject to statutory exception and to the provisions
of relevant collective licensing agreements, no reproduction of any part may take
place without the written permission of Cambridge University Press & Assessment.

This publication has been registered with Copyright Clearance Center, Inc.
For further information please contact the Copyright Clearance Center,
Salem, Massachusetts.

First published 2010
First paperback edition 2012

Single article reprints from this publication are available through
University Microfilms Inc., 300 North Zeeb Road, Ann Arbor, MI 48106

CODEN: MRSPDH

A catalogue record for this publication is available from the British Library

ISBN   978-1-605-11249-7   Hardback
ISBN   978-1-107-40673-5   Paperback

Cambridge University Press & Assessment has no responsibility for the persistence
or accuracy of URLs for external or third-party internet websites referred to in this
publication and does not guarantee that any content on such websites is, or will
remain, accurate or appropriate.

CONTENTS

*Invited Paper

*Invited Paper

vii

PREFACE

This volume contains papers presented at five symposia: Symposium KK, "Micro- and Nanofluidic Systems for Material Synthesis, Device Assembly, and Bioanalysis"; Symposium LL, "Directed Assembly and Self Assembly—From Synthesis to Device Applications"; Symposium NN, "Materials Exploiting Peptide and Protein Self Assembly—Toward Design Rules"; Symposium OO, "Hierarchical Self Assembly of Functional Materials—From Nanoscopic to Mesoscopic Length Scales"; and Symposium PP, "Interfacing Biomolecules and Functional (Nano) Materials." They were all held April 5–9 at the 2010 MRS Spring Meeting in San Francisco, California.

The integration of miniaturized components is essential to enable new functional capabilities for materials and devices. This integration can be achieved using conventional planar lithographic methods; however, cost-effective integration at nanometer size scales and in three dimensions remains challenging. Hence, there is a need to develop novel methods such as self-assembly, which is rapidly emerging as a new methodology to synthesize ordered structures from micro- and nanoscale building blocks.

The papers in this book are divided by the symposia at which they were presented, but have been collated in a single volume due to the numerous similarities and an overall theme of assembly and integration. The authors address the challenges in the cost-effective fabrication of miniaturized devices especially at small nano-sized scales and in three dimensions. These papers span a wide range of assembly methods including: molecular assembly methods with polypeptides and polynucleotides; rolling, wrinkling and self-folding of thin films; and self-assembly of lithographic structures using physical forces. Novel methods that seek to overcome challenges in enabling material and device designs for microfluidics and nanofluidics are included. Another area covered is that of biointerfaces and biomimetic materials with a focus on heterogeneous integration of materials from disparate material classes such as polymers, gels, and metallic & oxide-based nanoparticles.

Together, this collection of papers represents a comprehensive sampling of the state of the art of self-assembly and includes novel methods of device integration in micro- and nanotechnology. These are emerging fields and seek to transform electronics, medicine, optics, and the environment by enabling precisely nano- and micro-structured materials and devices on a hierarchy of length scales.

Carlos Martinez
Amit Goyal
Alberto Saiani
David Gracias
Rajesh Naik

November 2010

MATERIALS RESEARCH SOCIETY SYMPOSIUM PROCEEDINGS

Volume 1245 — Amorphous and Polycrystalline Thin-Film Silicon Science and Technology—2010, Q. Wang, B. Yan, C.C. Tsai, S. Higashi, A. Flewitt, 2010, ISBN 978-1-60511-222-0

Volume 1246 — Silicon Carbide 2010—Materials, Processing and Devices, S.E. Saddow, E.K. Sanchez, F. Zhao, M. Dudley, 2010, ISBN 978-1-60511-223-7

Volume 1247E —Solution Processing of Inorganic and Hybrid Materials for Electronics and Photonics, 2010, ISBN 978-1-60511-224-4

Volume 1248E —Plasmonic Materials and Metamaterials, J.A. Dionne, L.A. Sweatlock, G. Shvets, L.P. Lee, 2010, ISBN 978-1-60511-225-1

Volume 1249 — Advanced Interconnects and Chemical Mechanical Planarization for Micro- and Nanoelectronics, J.W. Bartha, C.L. Borst, D. DeNardis, H. Kim, A. Naeemi, A. Nelson, S.S. Papa Rao, H.W. Ro, D. Toma, 2010, ISBN 978-1-60511-226-8

Volume 1250 — Materials and Physics for Nonvolatile Memories II, C. Bonafos, Y. Fujisaki, P. Dimitrakis, E. Tokumitsu, 2010, ISBN 978-1-60511-227-5

Volume 1251E —Phase-Change Materials for Memory and Reconfigurable Electronics Applications, P. Fons, K. Campbell, B. Cheong, S. Raoux, M. Wuttig, 2010, ISBN 978-1-60511-228-2

Volume 1252— Materials and Devices for End-of-Roadmap and Beyond CMOS Scaling, A.C. Kummel, P. Majhi, I. Thayne, H. Watanabe, S. Ramanathan, S. Guha, J. Mannhart, 2010, ISBN 978-1-60511-229-9

Volume 1253 — Functional Materials and Nanostructures for Chemical and Biochemical Sensing, E. Comini, P. Gouma, G. Malliaras, L. Torsi, 2010, ISBN 978-1-60511-230-5

Volume 1254E —Recent Advances and New Discoveries in High-Temperature Superconductivity, S.H. Wee, V. Selvamanickam, Q. Jia, H. Hosono, H-H. Wen, 2010, ISBN 978-1-60511-231-2

Volume 1255E —Structure-Function Relations at Perovskite Surfaces and Interfaces, A.P. Baddorf, U. Diebold, D. Hesse, A. Rappe, N. Shibata, 2010, ISBN 978-1-60511-232-9

Volume 1256E —Functional Oxide Nanostructures and Heterostructures, 2010, ISBN 978-1-60511-233-6

Volume 1257 — Multifunctional Nanoparticle Systems—Coupled Behavior and Applications, Y. Bao, A.M. Dattelbaum, J.B. Tracy, Y. Yin, 2010, ISBN 978-1-60511-234-3

Volume 1258 — Low-Dimensional Functional Nanostructures—Fabrication, Characterization and Applications, H. Riel, W. Lee, M. Zacharias, M. McAlpine, T. Mayer , H. Fan, M. Knez, S. Wong, 2010, ISBN 978-1-60511-235-0

Volume 1259E —Graphene Materials and Devices, M. Chhowalla, 2010, ISBN 978-1-60511-236-7

Volume 1260 — Photovoltaics and Optoelectronics from Nanoparticles, M. Winterer, W.L. Gladfelter, D.R. Gamelin, S. Oda, 2010, ISBN 978-1-60511-237-4

Volume 1261E —Scanning Probe Microscopy—Frontiers in NanoBio Science, C. Durkan, 2010, ISBN 978-1-60511-238-1

Volume 1262 — In-Situ and Operando Probing of Energy Materials at Multiscale Down to Single Atomic Column—The Power of X-Rays, Neutrons and Electron Microscopy, C.M. Wang, N. de Jonge, R.E. Dunin-Borkowski, A. Braun, J-H. Guo, H. Schober, R.E. Winans, 2010, ISBN 978-1-60511-239-8

Volume 1263E —Computational Approaches to Materials for Energy, K. Kim, M. van Shilfgaarde, V. Ozolins, G. Ceder, V. Tomar, 2010, ISBN 978-1-60511-240-4

Volume 1264 — Basic Actinide Science and Materials for Nuclear Applications, J.K. Gibson, S.K. McCall, E.D. Bauer, L. Soderholm, T. Fanghaenel, R. Devanathan, A. Misra, C. Trautmann, B.D. Wirth, 2010, ISBN 978-1-60511-241-1

Volume 1265 — Scientific Basis for Nuclear Waste Management XXXIV, K.L. Smith, S. Kroeker, B. Uberuaga, K.R. Whittle, 2010, ISBN 978-1-60511-242-8

Volume 1266E —Solid-State Batteries, S-H. Lee, A. Hayashi, N. Dudney, K. Takada, 2010, ISBN 978-1-60511-243-5

Volume 1267 — Thermoelectric Materials 2010—Growth, Properties, Novel Characterization Methods and Applications, H.L. Tuller, J.D. Baniecki, G.J. Snyder, J.A. Malen, 2010, ISBN 978-1-60511-244-2

Volume 1268 — Defects in Inorganic Photovoltaic Materials, D. Friedman, M. Stavola, W. Walukiewicz, S. Zhang, 2010, ISBN 978-1-60511-245-9

Volume 1269E —Polymer Materials and Membranes for Energy Devices, A.M. Herring, J.B. Kerr, S.J. Hamrock, T.A. Zawodzinski, 2010, ISBN 978-1-60511-246-6

MATERIALS RESEARCH SOCIETY SYMPOSIUM PROCEEDINGS

Volume 1270 — Organic Photovoltaics and Related Electronics—From Excitons to Devices,
V.R. Bommisetty, N.S. Sariciftci, K. Narayan, G. Rumbles, P. Peumans, J. van de Lagemaat,
G. Dennler, S.E. Shaheen, 2010, ISBN 978-1-60511-247-3

Volume 1271E —Stretchable Electronics and Conformal Biointerfaces, S.P. Lacour, S. Bauer, J. Rogers,
B. Morrison, 2010, ISBN 978-1-60511-248-0

Volume 1272 — Integrated Miniaturized Materials—From Self-Assembly to Device Integration,
C.J. Martinez, J. Cabral, A. Fernandez-Nieves, S. Grego, A. Goyal, Q. Lin, J.J. Urban,
J.J. Watkins, A. Saiani, R. Callens, J.H. Collier, A. Donald, W. Murphy, D.H. Gracias,
B.A. Grzybowski, P.W.K. Rothemund, O.G. Schmidt, R.R. Naik, P.B. Messersmith,
M.M. Stevens, R.V. Ulijn, 2010, ISBN 978-1-60511-249-7

Volume 1273E —Evaporative Self Assembly of Polymers, Nanoparticles and DNA , B.A. Korgel, 2010,
ISBN 978-1-60511-250-3

Volume 1274 — Biological Materials and Structures in Physiologically Extreme Conditions and Disease,
M.J. Buehler, D. Kaplan, C.T. Lim, J. Spatz, 2010, ISBN 978-1-60511-251-0

Prior Materials Research Society Symposium Proceedings available by contacting Materials Research Society

**Micro- and Nanofluidic Systems for Material Synthesis,
Device Assembly, and Bioanalysis**

Mater. Res. Soc. Symp. Proc. Vol. 1272 © 2010 Materials Research Society   1272-KK03-05

Enhancement of on-chip bioassay efficiency with electrothermal effect

Kuan-Rong Huang, Jeng-Shian Chang, Sheng D. Chao, and Kuang-Chong Wu

Institute of Applied Mechanics, National Taiwan University, Taipei 106, Taiwan

ABSTRACT

We have performed the finite element simulations to study the binding reaction kinetics of the analyte-ligand protein pairs, C-reactive protein (CRP) and anti-CRP, in a reaction chamber of a biosensor. For diffusion limited reactions, diffusion boundary layers often develop on the reaction surface, thus hindering the reaction. To enhance the efficiency of a biosensor, a non-uniform AC electric field is applied to induce the electrothermal force which stirs the flow field. Biosensors with different arrangements of the electrode pairs and the reaction surface are designed to study the effects of geometric configurations on the binding efficiency. The maximum initial slope of the binding curve can be 6.94 times of the field-free value in the association phase, under an AC field of 15 V*rms* and an operating frequency of 100 *kHz*. With the electrothermal effect, it is possible to use a slower flow and save much sample consumption without sacrificing the performance of a biosensor. Several design factors not studied in our previous works such as the thermal boundary conditions are discussed.

INTRODUCTION

Immunoassay is a versatile biomedical diagnostic tool and provides a way for highly sensitive and precise detection of biological agents [1, 2]. The possibility of fabricating and integrating micro-sized biosensors with multiple functions has led to the idea of performing real-time monitoring or diagnostics on a portable lab-on-a-chip. Most biosensors employ the same kinetics of specific binding of analytes, such as C-reactive protein (CRP) [3], and immobilized ligands, such as anti-CRP. When the flow rate is fixed, the required experiment time of a specific bio-molecular recognition usually depends on its Damköhler number (*Da* number), which is the ratio of reaction velocity (i.e., product of the association rate constant and the initial concentration of the ligand) to the diffusion velocity (i.e., ratio of the diffusion coefficient of the analyte in the buffer flow to the height of micro-channel) [4]. When the *Da* number is greater than unity, the whole reaction is restrained by mass-transport process called diffusion limited reaction. This usually occurs in conditions of slow flow velocity in micro-channel, small diffusion coefficient of the analyte, and high association rate constant of reaction.

If the whole reaction is restrained by mass-transport, a diffusion boundary layer is often formed on the reaction surface [5, 6]. The formation of such layers would limit the response-time and the overall performance of a biosensor. In practice, it often takes hours to complete a detection cycle, which is the main technical problem to be solved. It is promising in our previous study [6] and the study of Sigurdson et al. [7] that several fabrication parameters can be tuned to improve the performance of a biosensor. From previous studies [8-19], the induced electrothermal force by the AC electric field can create a vortex field, called electrothermal flow, which will enhance the transport of the analyte to the reaction surface, reduce the thickness of the diffusion boundary layer and significantly increase the reaction rate to accelerate both the association and dissociation processes. However, these simulations were done only for two dimensional cases, limiting the spatial factors for practical designing. In this study, we will perform three dimensional simulations and study the effect of changing the relative geometrical locations of the electrodes and reaction surface on the biosensor performance. The sample of CRP pairs in a biosensor immunoassay will be used to predict the surface concentration of the analyte-ligand complex versus time relationship.

THEORY

The details of the methodology used for the present 3D simulations are similar to our previous 2D simulations [6] and will be summarized briefly. This electrothermal force is given by [8]:

$$\vec{F_E} = -\frac{1}{2}\left[(\frac{\nabla\sigma}{\sigma} - \frac{\nabla\varepsilon}{\varepsilon})\cdot\vec{E}\frac{\varepsilon\vec{E}}{1+(\omega\tau)^2} + \frac{1}{2}\left|\vec{E}\right|^2\nabla\varepsilon\right]$$

(1)

where $\tau = \varepsilon/\sigma$ is the charge relaxation time, ω is the angular frequency of the electric field $\vec{E}$, The electrostatic field $\vec{E}$ is related to the electrical potential Φ by :

$$\vec{E} = -\nabla\Phi$$

(2)

where Φ satisfies the Laplace's equation [20]

$$\nabla^2\Phi = 0$$

(3)

The following energy balance equation must be solved [21].

$$\rho c_p\frac{\partial T}{\partial t} + \rho c_p\vec{V}\cdot\nabla T = k\nabla^2 T + \sigma\left|\vec{E}\right|^2$$

(4)

where ρ, c_p, κ and $\vec{V}$ are the density, specific heat, thermal conductivity, and velocity of the fluid, respectively. In this work it is assumed that the fluid is incompressible so that

$$\nabla\cdot\vec{V} = 0$$

$$\rho(\frac{\partial\vec{V}}{\partial t} + \vec{V}\cdot\nabla\vec{V}) - \eta\nabla^2\vec{V} + \nabla p = \vec{F_E}$$

(5)

where η is the dynamic viscosity of the fluid and p is the pressure. Transport of analytes to and from the reaction surface is assumed to be described by the Fick's second law with convective terms:

$$\frac{\partial[A]}{\partial t} + \vec{V}\cdot\nabla[A] = D\nabla^2[A]$$

(6)

where $[A]$ is the concentration of analyte, and D is the diffusion coefficient of analyte in the fluid. The reaction between immobilized ligand and analyte is assumed to follow the first order Langmuir adsorption model [5, 22]. During the reaction, the analyte-ligand complex $[AB]$ increases as a function of time according to the reaction rate:

$$\frac{\partial[AB]}{\partial t} = k_a[A]_{surface}\left\{[B_0] - [AB]\right\} - k_d[AB]$$

(7)

where $[A]_{surface}$ is the concentration of the analyte at the reaction surface by mass-transport, $[B_0]$ is the surface concentration of the ligand, and $[AB]$ is the surface concentration of the analyte-ligand complex. Sketch of the 3-D model of micro-channel is shown in Figure 1. In this work, the dimensions of the micro-channel and the reaction surface are $1000\mu m \times 500\mu m \times 100\mu m$ and

4

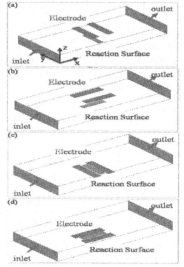

$200 \mu m \times 40 \mu m \times 3 \mu m$, respectively. The dimensions of each electrode are 200 μm in length, 60 μm in width. The interval between the two electrodes is fixed at 60 μm.

DISCUSSION

Fig. 1. The four arrangements of the electrodes and reaction surface.

The arrangements of the electrode pair and reaction surface

Figure 1 shows the four types of biosensors with different arrangements of the electrodes and reaction surface considered in this study. The applied voltage is 15 V and the inlet flow velocity is 100 μm/s. Table 1 lists the values of the initial slope and also the enhancement factors, defined as the ratio of the slope of binding reaction with applying electrothermal effect to that without applying electrothermal effect. Our previous study [6] revealed that the longer the length of the reaction surface along the flow direction is, the larger the diffusion boundary layer grows. The existence of the diffusion boundary layer on the reaction surface hinders the binding reaction of analyte-ligand pairs. Since Type-2/Type-4 biosensor has the longer dimension along the flow direction than Type-1/Type-3 biosensor does, the former is hindered more on the binding reaction than the latter. However, such geometrically caused hindrance can be removed by applying the external electrothermal field. The binding curve of Type-2 biosensor coincides almost with that of Type-1 biosensor after the electrothermal field is applied. Similar situation occurs for the Type-3 and Type-4 biosensors. It clearly suggests that the reaction surface should be put together with the electrodes at the same wall of the micro-channel to get better performance. Therefore the following simulations will focus on Type-3 and Type-4 biosensors.

Table 1 The initial slopes and the enhancement factors for the four biosensors.

Type of biosensor	Initial slope (association) $\times 10^{-11}$	Initial slope (dissociation) $\times 10^{-11}$	Enhancement factor (association)	Enhancement factor (dissociation)
Type-1	2.30	-1.72	1.06	1.06
Type-2	2.21	-1.66	1.45	1.38
Type-3	4.42	-2.79	2.04	1.71
Type-4	4.51	-2.81	2.95	2.33

The thermal boundary conditions

5

Four cases of thermal boundary setting to the electrodes, and the top and bottom walls are considered in the simulation. For Case-1, the electrodes are controlled at 300 K, but both the top and bottom walls of the micro-channel are insulated. This is the case adopted in the above simulations. For Case-2, both the temperature of the electrodes and the bottom wall of the micro-channel are maintained at 300K but the top wall is insulated. For Case-3, both the temperature of the electrodes and the top wall of the micro-channel are maintained at 300K but the bottom wall is insulated. Finally, for Case-4, both the electrodes and the walls are all maintained at 300K. Moreover, the applied voltage is 15 V and the inlet flow velocity is 100 μm/s.

Table 2 Initial slopes and enhancement factors of the CRP binding reactions of Type-3/4 biosensor for the four cases of thermal boundary conditions.

Temperature boundary conditions	Initial slope (association) $\times 10^{-11}$	Enhancement factor (association)
Case-1 (electrodes at 300K)	4.43/4.52	2.03/2.93
Case-2 (electrodes and bottom wall at 300K)	4.05/4.37	1.86/2.84
Case-3 (electrodes and top wall at 300K)	3.80/4.04	1.75/2.62
Case-4 (electrodes, bottom wall and top wall at 300K)	3.73/4.08	1.72/2.65

Table 2 lists both the initial slopes and the enhancement factors for the association reaction for the Type-3/4 biosensor with the four different cases of temperature control. It is seen that the Case-1 temperature boundary setting produces the larger temperature rise and gradient around the reaction surface than the other three cases and turns out to yield the best performance among these four cases of the boundary temperature setting in terms of the initial slopes and enhancement of binding reaction,.

Effect of the inlet flow velocity

Table 3 shows the initial slopes and the enhancement factors in the association phase of the CRP binding reaction with various inlet flow velocities. It can be seen that higher flow velocity causes larger convective mass transport of the analyte and hence larger initial slope of the binding reaction. Nevertheless, the enhancement factor decreases monotonically as the inlet flow velocity increases. It means that the electrothermal effect is more efficient at lower inlet flow velocity than at higher velocity. For example, the enhancement factor is 6.94 for Type -4 biosensor when the inlet flow velocity is decreased to 5 μm/s, but only 1.25 for Type-3 biosensor when the inlet flow velocity is increased to 500 μm/s. Although the binding reaction rate for higher inlet flow velocity is still faster than the one for lower inlet flow velocity, the wasteful consumption of the sample is not worthy since most of them are carried away by the flow. Consider the Type-4 biosensor, with the inlet flow velocity and bulk concentration being 100 $\mu m/s$ and 6.4 nM, respectively. The saturation times t_{sat} for the binding reaction (not shown here) are about 1500s and 400s for 0 V and 15 V applied AC voltage, respectively. The

6

ratio of the bound analyte on the reaction surface to the total amount of the analyte supplied when the reaction is saturated can be computed as

$$\frac{[A]_{sat} \times Area_{reaction\ surface}}{[A]_{bulk} \times (A_{inlet} \times inlet\ velocity \times t_{sat})} \approx \begin{cases} 0.17\% & for\ 0\ V \\ 0.63\% & for\ 15\ V \end{cases}$$

This simple calculation shows that the absorbed analyte is only less than 1 % of the total supplied samples and most of the samples go to waste. This ratio becomes even smaller when the inlet flow velocity becomes higher, say $500\ \mu m/s$.

Table 3 Initial slopes and enhancement factors in the association phase of Type-3/4 biosensor

with Case-1 temperature condition for various inlet flow velocities.

w Velocity ($\mu m/s$)	Initial Slope $\times 10^{-11}$ (association), $0\ V_{rms}$	Initial Slope $\times 10^{-11}$ (association), $15\ V_{rms}$	Enhancement Factor (association)
5	0.14/0.14	0.77/0.98	5.41/6.94
10	0.63/0.52	2.20/1.99	3.46/3.81
30	1.34/1.01	3.99/3.82	2.97/3.77
50	1.68/1.23	4.32/4.34	2.57/3.53
100	2.18/1.53	4.42/4.51	2.04/2.95
300	3.10/2.13	4.48/4.61	1.45/2.17
500	3.59/2.46	4.49/4.67	1.25/1.90

In this respect, the advantage of applying electrothermal field is outstanding. It can efficiently raise the binding speed at lower inlet flow velocity to save the sample consumption without sacrificing too much the performance of a biosensor. We also observe that when no electrothermal field is applied, the performances of Type-3 and Type-4 biosensors are remarkably distinguished from each other. In fact, Type-3 biosensor has shorter dimension along the flow direction than Type-4 biosensor does, and hence has a smaller diffusion boundary layer zone than Type-4 biosensor does. The diffusion boundary layer hinders the binding reaction, as revealed in our previous study [6]. Thus Type-3 biosensor has better performance in binding reaction than Type-4 biosensor does. However, when the electrothermal field is applied, the diffusion boundary layers are almost completely eliminated for both Type-3 and Type-4 biosensors, and so this geometric effect to hinder the reaction rate is removed, and the binding reaction curves of Type-3 and Type-4 biosensors are quite close to each other now.

CONCLUSIONS

We study the electrothermal effect on the binding efficiency by performing three dimensional simulations on the immunoassay biosensor using the finite element analysis software, COMSOL Multiphysics [TM] [23]. Our results indicate that the drawback of a longer reaction surface along the flow direction can be removed by applying the electrothermal force. A benefit of employing the electrothermal effect is that slower flow velocity in an immunoassay can be used to save sample consumption. Different boundary temperature controls can influence the binding rate significantly.

ACKNOWLEDGMENTS

This research was supported by the National Science Council in Taiwan through NSC 97-2221-E-002-017-MY3 and National Taiwan University CQSE 97R0066-69. We thank the NCHC in Taiwan for providing computing resources.

REFERENCES

[1] R. S. Yallow, and S. Berson, *Nature*, 184 (1959) 1648.
[2] P. A. Auroux, D. Iossifidis, D. R. Reyes, and A. Manz, *Anal. Chem.*, 74 (2002) 2637.
[3] W. S. Tillet, and T. Francis, *J. Exp. Med.*, 52 (1930) 561.
[4] W. M. Deen, *Analysis of Transport Phenomena*, Oxford University Press, New York, 1998.
[5] D. B. Hibbert, and J. J. Gooding, *Langmuir*, 18 (2002) 1770.
[6] C. K. Yang, J. S. Chang, S. D. Chao, and K. C. Wu, *J. App. Phys*, 103 (2008) 084702.
[7] M. Sigurdson, D. Wang, and C. D. Meinhart, *Lab on a chip*, 5 (2005) 1366.
[8] Ramos, H. Morgan, and A. Castellanos, *J.Phys.D:Appl.Phys.*, 31 (1998) 2338.
[9] R. Pethig, *Crit. Revs. Biotechnol.*,16 (1996) 331.
[10] Ramos, A. Gonzalez, N. Green, A. Castellanos, and H. Morgan, *Phys Rev.*, 61-4 (2000) 4019.
[11] N. Green, A. Ramos, A. Gonzalez, H. Morgan, and A. Castellanos, *Phys Rev.*, 61-4 (2000), 4011.
[12] C. Meinhart, D. Wang, and K. Turner, *Biomedical Microdevices*, 5-2 (2003) 139.
[13] X. B. Wang, Y. Huang, P. R. C. Gascoyne, and F. F. Becker, *IEEE Trans. Ind. Appl.*, 33 (1997) 660.
[14] H. Morgan, M. P. Hughes, and N. G. Green, *Biophysical J.*, 77 (1999) 516.
[15] M. Washizu, and S. Suzuki, *IEEE Transaction on Industry Application*, 30 (1994) 835.
[16] D. Wang, M. Sigurdson, and C. D. Meinhart, *Experiments in Fluid*, 38 (2005) 1.
[17] H. C. Feldman, M. Sigurdson, and C. D. Meinhart, *Lab on a chip*, 7 (2007) 1553.
[18] K. R. Huang, J. S. Chang, S. D. Chao, K. C. Wu, C. K. Yang, C. Y. Lai, S. H. Chen, *J. App. Phys.* 104 (2008), 064702.
[19] C. K. Yang, J. S. Chang, S. D. Chao, and K. C. Wu, *App. Phys. Lett.*,. 91 (2007), 113904.
[20] J. A. Stratton, *Electromagnetic Theory*, McGraw Hill, New York, 1941.
[21] L. D. Landau and E. M. Lifshitz, *Fluid Mechanics*, Pergamon, Oxford, 1959.
[22] Langmuir, *J. Am. Chem. Soc.*, 40 (1918) 1361.
[23] Comsol Multiphysics Version 3.4, COMSOL Ltd., Stokhelm, Sweden.

Mater. Res. Soc. Symp. Proc. Vol. 1272 © 2010 Materials Research Society   1272-KK05-01

Experimental characterization of the whipping instability of charged microjets in liquid baths

G. Riboux[1], A. G. Marín[1], A. Barrero[1], A. Fernández-Nieves[2] & I. G. Loscertales[3]

[1] Escuela Superior de Ingenieros, Universidad de Sevilla, Seville, Spain
[2] School of Physics, Georgia Institute of Technology, GA, USA.
[4] Escuela de Ingenierías, Universidad de Málaga, Málaga, Spain

ABSTRACT

Capillary liquid flows have shown their ability to generate micro and nano-structures which can be used to synthesize material in the micro or nanometric size range. For instance, electrified capillary liquid jets issued from a Taylor are broadly used to spin micro and nanofibers when the liquid consists of a polymer solution or melt, a process termed electrospinning. In this process, the electrified capillary jet may develop a non-axisymmetric instability, usually referred to as whipping instability, which very efficiently transforms electric energy into stretching energy, thus leading to the formation of extremely thin polymer fibers. Even though non axisymmetric instabilities of electrified jets were first investigated some decades ago, the existing theoretical models provide a qualitative understanding of the phenomenon but none of them is accurate enough when compared with experimental results. This whipping instability usually manifests itself as fast and violent lateral motion of the charged jet, which makes it difficult its characterization in the laboratory. However, this instability also develops when electrospinning is performed within a liquid bath instead of air. Although it is essentially the same phenomenon, the frequency of the whipping oscillations is much slower in the former case than in the latter, thus allowing detailed experimental characterization of the whipping instability. Furthermore, since the outer fluid is a liquid, its density and viscosity may now be used to influence the dynamics of the electrified capillary jet. In this work we present and rationalize the experimental data collecting the influence of the main parameters on the whipping characteristics of the electrified jet (frequency, amplitude, etc.).

INTRODUCTION

The electro-hydrodynamic (EHD) stretching of a liquid meniscus is among the most popular and versatile techniques to form extremely thin electrified capillary jets that may be used to synthesize various types of micro or nanoparticles [1]. This EHD technique is usually split into those used to produce spherical-like particles, termed electrosprays, and those used to form fiber-like particles, called electrospinning. In either case, an electrified conical meniscus, called Taylor cone, is formed at the tip of a capillary tube; from the tip of that cone, a thin electrified jet of liquid is issued. The nature of this phenomenon renders the charged jet unstable under axisymmetric (varicose) and non-axisymmetric perturbations. The former ones tend to break the jet into droplets with diameter of the order of the jet, similarly to what happens to uncharged jet break up, and so the processing of these airborne micro or nanodroplets gives rise to micro or nanoparticles. The non-axisymmetric instability, which do not appear in uncharged jets, manifest itself spontaneously as very fast and violent spiral-like and lateral motions of the jet; in this case, the electric energy is efficiently transferred into stretching energy and so, if the jet survives this stressing motion, its diameter is dramatically thinned. This type of instability is called whipping instability and is responsible for the formation of nanofibers from polymer solutions or melts.

Since both types of instabilities are competing, it becomes naturally interesting to look for the parameters that determine which one grows faster in order to adjust the experimental parameters to produce either nanoparticles or nanofibers. Even though the work on electrospray and electrospinning has grown dramatically during the last decade, most of the literature on electrospinning deals with phenomenological studies testing the electrospinnability and properties of different polymer solutions, and only few works are devoted to theoretically predicting the behavior of the charged jet [2-4]. Some of the theoretical works qualitatively predict the dominance of either instability, or the expected final diameter of the electrospun fibers, but seldom describe in a detailed manner the dynamics of the whipping. It then seems appropriate to devote some effort to characterize the whipping. Unfortunately, this task becomes rather difficult mainly due to the very short characteristic times of this violent motion, which makes it extremely difficult to record or to photograph its evolution.

Recently, Barrero's group [5] operated an electrospray in steady cone-jet mode inside a dielectric liquid medium. Although electrosprays within a host liquid medium are ruled by the same physical principles than those operated in air, there are, however, some differences in their behaviors. For instance, the oscillations of the whipping instability are much less violent when the process is run inside a dielectric liquid than in air. This effect, due to the much larger value of the inertia of the outer fluid, results in jet oscillations at frequencies several orders of magnitude lower than those typically found in electrospinning in air [3, 6, 7]. Taking advantage of this damping effect, we have carried out an experimental characterization of the whipping instability of a charged capillary jet issued from an electrospray of glycerin operated within hexane (a dielectric liquid bath). To do so we have measured the whipping frequency, amplitude and wavelength in terms of the governing parameters, flow rate and electrical field.

EXPERIMENTAL

The experimental setup is entirely similar to that reported in previous works on electrosprays inside liquid media [4, 8]. It consists of an open tank of Plexiglas with a $38 \times 32 \, mm^2$ cross-section and 180 mm high, see figure 1a. Two of the walls of the tank were made of glass to allow visualization. The tank was filled with hexane, whose viscosity, density and relative electrical permittivity were, respectively, $\mu_e = 0.3 \, mPa.s$, $\rho_e = 660 \, kg.m^{-3}$ and $\frac{\epsilon_e}{\epsilon_o} = 1.89$, where ϵ_e is the electrical permittivity of the hexane and ϵ_o is the vacuum permittivity. A round electrode-collector of $23 \, mm$ of diameter connected to electrical ground was located at the bottom of the tank. A metallic needle of inner radius $a = 115 \, \mu m$ immersed in the hexane was positioned at a fixed distance $H = 27 \, mm$ above the collector. Glycerin was injected through the needle at a flow rate Q; a Harvard PHD 4400 programmable syringe pump with a Hamilton Gastight glass syringe of $32.6 \, mm$ in diameter was used to control the flow rate. The physical properties of the glycerin were: density $\rho = 1250 \, kg.m^{-3}$, dynamic viscosity $\mu = 1280 \, mPa.s$, electrical conductivity $K = 1.7 \times 10^{-6} \, S.m^{-1}$ and electrical permittivity $\frac{\epsilon}{\epsilon_o} = 43$.

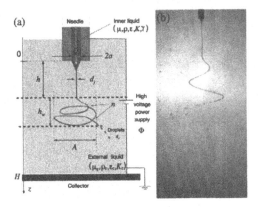

Figure 1: (a) Sketch of the experimental set up. (b) An example picture of the whipping instability ($Q = 1.0 \frac{ml}{h}$, $\Phi = 2.25\ kV$).

Surface tension of the glycerin-hexane pair was measured using the pendant drop method. The measured value was $\gamma = 28.3\ mN. m^{-1}$. The metallic needle was connected to a Bertan 205B-10R high voltage power supply to impose an electric field between the collector and the needle exit (see Fig. 1a). The whipping motion of the jet was captured with a Photron FASTCAM 1024 PCI high-speed video camera. It was recorded at an acquisition frequency equal to 2000 frames per second. We used a floodlight facing the high-speed video camera to obtain a good contrast of the jet image. The dimensions in all pictures were 1024×512 pixels while the resolution was within a range of $8 - 16\ \mu m$. Figure 1b presents an example of a digital image of the jet path. To analyze the images of the jet motion, we have used the image processing software *ImageJ* together with algorithms developed in house. Examples of digital jet paths are presented in figure 2a. This figure shows two digital images of the jet path. The images were captured at two different times separated by an interval of 29 milliseconds in this case. Since the two images perfectly superpose on top of each other, one may conclude that the whipping mode is periodic for this set of values of the governing parameters ($Q = 1.0\ ml/h$, $\Phi = 2.25\ kV$); the period of the whipping oscillations being 29 milliseconds. The value of the whipping period may be measured with an accuracy of ±1 frames, which corresponds to 0.5 ms. Figure 2b shows two images of the jet path taken in a lapse of time of a semi-period (14.5 ms); the superposition of the two digital images permits the measurement of the wavelength λ_i of the whipping oscillation in this periodic regime. It should be noted that as a consequence of the electric field, the wavelength λ_i of the oscillation slightly varies along the z-axis. Therefore, we define a mean value λ of the wavelength by averaging the values of the measured wavelength λ_i of each jet loop along the z-axis.

The envelope of the whipping jet paths can be also obtained by overlapping the jet images captured at different times. Figure 2c shows the result of overlapping 300 images of the whipping jet. This overlapping allows an easy digitalization of the envelope of the jet paths and hence the definition of the envelope's contour as a function of the distance z to the needle. In figure 2c, we have plotted for comparison both the binary image of the envelope of the jet path

11

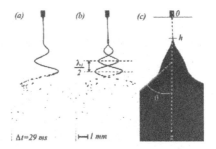

Figure 2: Images of the trajectories of the whipping jet. (a) Images of the lateral instability taken at two instants elapsed by a complete period. (b) Wavelength measurements at a half period. (c) Whipping envelope with θ the maximal angle of the droplet trajectories ($Q = 1.0\,\frac{ml}{h}$, $\Phi = 2.25\ kV$).

(black region) and the overlapping of the binary jet paths (grey zone). Since the whipping envelope is symmetrical, it is possible to obtain an averaged envelope meridional curve as a function of the distance z to the needle from data of the binary images and MATLAB algorithms. It should be noted that other possible choices for the envelope detection, as the use of video cameras with long time exposure shots yielded less clean and neat results.

DIMENSIONLESS PARAMETERS: SCALING

The experimental characterization of the whipping instability of an electrified liquid jet evolving within either a gas atmosphere or a bath of a different liquid is a complex problem due to the large number of physical parameters that enter in the phenomenon. In the general case of a microjet subjected to a strong electric field, the relevant parameters are those associated with the physical properties of the two fluids (viscosity μ, density ρ, electrical permittivity ϵ and conductivity K of each fluid and the surface tension γ of the fluid couple), the geometrical characteristics of the experimental set up (in our case, the radius of the capillary a and capillary to collector distance H) and, finally, the electrospray control parameters (the flow rate Q and the applied electrical voltage Φ). All these add up to 13 parameters. One must select four of them to obtain a characteristic time, length, mass and charge, which reduces the number of independent dimensionless parameters down to 9. Nonetheless, in our experimental case, the effect of the viscosity of the outer fluid (hexane) can be neglected since the hexane-glycerin viscosity ratio is of the order of 10^{-4}. Also, since hexane acts as an insulator, its electric conductivity also drops out of the picture. Furthermore, the Reynolds number $\rho Q/(\mu a)$ of the jet results to be of the order of 10^{-3}, so we neglected the inertia of the inner liquid. Accordingly, the dynamic of a charged viscous capillary jet formulated in non-dimensional form is governed by six dimensionless parameters. Using the viscosity of the glycerin, the surface tension, the capillary radius and the electrical permittivity of the outer fluid as dimensional scales of the problem, we arrive at the following dimensionless parameters of the problem: geometrical length ratio H/a, glycerin to hexane electrical permittivity ratio $\beta = \frac{\epsilon}{\epsilon_e}$, the dimensionless number $S = \rho_e \gamma a / \mu^2$ and

$$Ca = \frac{\mu Q}{\gamma a^2}, \; \mathcal{B} = \frac{\epsilon_e \beta \Phi^2}{\gamma a}, \; \mathcal{T} = \frac{\mu K a}{\epsilon \gamma}. \tag{1}$$

Observe that in our experiment, the dimensionless number $S = 1.3 \times 10^{-3}$, which is a Reynolds number for the outer fluid, is extremely small indicating that the relative importance in the dynamic of the outer fluid is rather small. The three dimensionless numbers in equation (1), which were already introduced by Higuera [9], are, respectively, the capillary number, the electrical Bond number and a dimensionless time. The capillary number Ca compares the fluid velocity Q/a^2 at the end of the capillary to the viscous-capillary velocity $Vo = \gamma/\mu$. The electrical Bond number $\mathcal{B}$ compares the electrical to capillary pressure. Finally, $\mathcal{T}$ is the ratio between the convection time $a\mu/\gamma$ and the electrical relaxation time ϵ/K.

In this work we fixed the fluids (except the conductivity of glycerin), the geometrical parameter $H/a = 235$ and $\beta = 23$, so the governing parameters are just Ca, $\mathcal{B}$ and $\mathcal{T}$. They may be independently varied through the flow rate Q, the applied electrical voltage Φ and by changing the electrical conductivity of glycerin, respectively. The ranges were $Ca = 0.2 - 8.5$, $\mathcal{B} = 300 - 1870$ and $\mathcal{T} = 19 - 27$.

PRELIMINARY OBSERVATIONS

Let us comment some preliminary observations before presenting the experimental results. Figure 3 shows the whipping-map for the fluid pair, glycerin-hexane. The whipping-map yields the behavior of the jet as a function of the control parameters (Q,Φ) or, equivalently, $(Ca, \mathcal{B})$. This type of graphic was previously introduced in [6, 7] for an electrified jet of either glycerin or of a mixture of PEO-water evolving in air; they called it operating diagram. Their diagrams are more general than ours since their also distinguish other instability modes of the jet.

Here, we report four sets of measurements whose results are collected in figure 3. The first set of eight measurements was taken at different values of the capillary number Ca while we kept $\mathcal{B} = 0$; this set of measurements, labeled as set A, is represented by square symbols and numbered from 1 to 8. In the range of capillary numbers considered, the experimental set up produces glycerin drops with radius of the order of 10 times the capillary radius; the frequency of the droplet formation increases with the capillary number. A long and thin jet was formed before the detachment of the droplet; the length of the jet just at the break-up process increases linearly with the capillary number Ca. At the jet breakup, one can observe the formation of a neck just behind the droplet.

The second set of measurements was taken at a four given values of the electrical Bond number $\mathcal{B}$ while the value of the capillary number was kept constant ($Ca = 1$). This set of four measurements, labeled as B in figure 3, is represented by diamond symbols and numbered from 1 to 4. We observed clearly the transition from a dripping mode (image 3, $\mathcal{B} = 260$) to a whipping mode (image 4, $\mathcal{B} = 470$). Therefore, one may define a value of the Bond number $\mathcal{B} = \mathcal{B}_{min}(Ca)$ such that for values of the Bond number $\mathcal{B} \leq \mathcal{B}_{min}$ the jet is just formed as a result of dripping while it stars whipping when $\mathcal{B} \geq \mathcal{B}_{min}$. In this set of measurements, we have

13

observed that the diameter of the drops decreased linearly when the electrical Bond number increased.

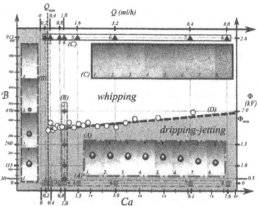

Figure 3: Whipping jet map for the glycerin-hexane couple, ($\mathcal{T} = 19$).

The third set of eight measurements was carried out keeping the electrical Bond number at a given value, $\mathcal{B}$ =915 while we varied the capillary number Ca in the range 0.2 to 7.6. This set is labeled as C and the experimental measurements are represented by triangular symbols and numbered from 1 to 8. In this case, we observed a transition from a jetting mode to a whipping mode (transition between image 1 and 2). Thus, and similarly to the previous case, there exists a capillary number (Ca_{min}) separating a jetting mode region from the one where the whipping instability develops. The jet diameter and the whipping amplitude increase with Ca. We have also observed a spatial destabilization of the jet motion when the capillary number increases. For $Ca = 1.5$, see (image C5), the spatial behavior of the jet appears more chaotic than for smaller values of Ca. (d) Finally we have determined the transition between the dripping mode to the whipping mode in the parametric space ($Ca, \mathcal{B}$) using a high-speed camera. The experimental measurements of this set, labeled as D in figure 3, are represented by empty circles. In the range of $Ca = 0.3 - 6.2$, the minimum electrical Bond number increases linearly with the capillary number. Thus, for the fluid pair, glycerin-hexane, the dripping-whipping transition boundary seems to evolve as $\mathcal{B}_{min} = 15Ca + 340$. Regarding the jetting-whipping transition, we have found that Ca_{min} has constant value, close to 0.3, which is completely independent of the range of values of $\mathcal{B}$ we have explored. In conclusion, the whipping instability appears only when the couple of parameters Ca and $\mathcal{B}$ are larger than $Ca_{min} \approx 0.3$ and $\mathcal{B}_{min} = 15Ca + 340$. Note that the whipping mode develops within the parametric region represented by the white domain in figure 3, while no whipping instability develops into the region in grey color.

WHIPPING MODE CHARACTERIZATION

This section is devoted to the experimental analysis of the jet whipping. We have observed whipping motion in all experimental cases since Ca and $\mathcal{B}$ were larger than the minimum values ($Ca_{min}, \mathcal{B}_{min}$). Therefore, all the experiments lay within the white whipping

14

region in figure 3. We have focused our experimental study on the jet whipping frequency, wavelength and the whipping envelope, as a function of Ca and $\mathcal{B}$.

Frequency and wavelength of the whipping instability

In figure 4, we have plotted the dimensionless whipping wavelength λ/a, line 1, the dimensionless whipping frequency fa/V_o, line 2, and the whipping phase velocity $f\lambda/V_o$, line 3 as a function of Ca, see figures (a), and $\mathcal{B}$, figures (b). Note that because the path of the jet motion may be non-periodic, the wavelength λ was only measurable in the spatially periodic part of the jet trajectory.

We have found that the wavelength λ depends on the dimensionless time ratio $\mathcal{T}$ and that it can be expressed as $\lambda \sim \lambda_o(\mathcal{T}/\mathcal{T}_o)$ with $\lambda_o = 12a = 1.4\,mm$ while the frequency can be expressed as $f \sim f_o(\mathcal{T}/\mathcal{T}_o)^{-1}$ with $f_o = V_o/3a = 63\,Hz$; f_o and λ_o are marked by dashed lines and the two plain lines point out the limits of the interval of $\pm15\%$ around these two values. Within the range $0.3 \leq Ca \leq 2.8$ the dimensionless ratios $(\lambda/a)(\mathcal{T}_o/\mathcal{T})$ and $(fa/V_o)(\mathcal{T}/\mathcal{T}_o)$ take constant values. For comparison, the values of the frequency reported in [6] in their experiments with a PEO (Poly-Ethylene Oxide)-water mixture in air has been also plotted (white circle symbols in figure 4 (2a)). In their experiment, these authors fed a constant volumetric flow rate $Q = 0.2\,ml/min$ of a PEO-water mixture through a stainless steel capillary tube of inner diameter $1\,mm$. The electric field in their experimental set-up was $E = 1.11\,kV/cm$ on a distance of the order of $10^{-2}\,m$. These experimental parameters correspond to $Ca = 0.4$, $\mathcal{T} = 2.4 \times 10^5$ and $\mathcal{B} \sim o(10 - 100)$. Resorting to long exposure times, these authors experimentally recorded the whipping envelope of this couple of fluids and fitted the experimental envelope by using an exponential law for the whipping amplitude of the type $A(z)/Ao \sim exp[f\,z/U_o]$; U_o and $f = 0.014 \pm 0.002\,s^{-1}$ being respectively the average fluid velocity at the capillary end and the frequency. Note that the experimental frequency reported in [6] is in good agreement with our results, so that, the expression $f_o(\mathcal{T}/\mathcal{T}_o)$ may be used with confidence for the PEO-water mixture in spite of the very large difference in the electrical conductivity with respect to that of the glycerin in our experiments. Finally, observe that for $Ca > 2.8$, the wavelength and the frequency of the whipping slightly increases and decreases, respectively, with the capillary number. From this limit, $Ca > 2.8$, we have observed that the the jet path is more unstable in time and space, and corresponds practically to a chaotic regime. As a consequence of the simultaneous increasing of the wavelength and decreasing of the frequency, the phase velocity of the whipping jet, defined as the product $f\lambda$, is constant, as shown in figure 4 (3a), and equal to $4V_o$ with an error of $\pm15\%$. Contrarily, both f and λ strongly vary with the electrical Bond number, following power laws of the type $\mathcal{B}^{3/2}$ and $\mathcal{B}^{-3/2}$, respectively, figure 4 (1b&2b). However, the whipping phase velocity $f\lambda$ is also practically independent of the electrical Bond number within the error band of $\pm15\%$, figure 4 (3b). In conclusion, the phase velocity of the jet whipping only depends on the viscous-capillary scale velocity V_o as $f\lambda = 4V_o$, which is $8.8\,cm/s$ for the glycerin-hexane couple. The wavelength λ can be expressed as,

$$\lambda = \lambda_o(\mathcal{B}/\mathcal{B}_o)^{-3/2}\,(\mathcal{T}/\mathcal{T}_o),\tag{2}$$

and the frequency f of the jet whipping can be expressed as,

$$f = f_o(\mathcal{B}/\mathcal{B}_o)^{3/2}\,(\mathcal{T}/\mathcal{T}_o)^{-1},\tag{3}$$

with $\mathcal{B}_o = 920$, $\mathcal{T}_o = 23$, $\lambda_o = 12a$ and $f_o = V_o/3a$. These results confirm the destabilizing effect of the electrical field on the jet trajectory as it has been typically observed in electrospinning experiments [6, 7], or in the experiments carried out by Taylor [10]. In fact, an increase in the applied voltage increases both the electrical field and the charge density of the jet. The jet becomes more unstable, increasing its whipping frequency, as it was commented by Hohman et al. (2001b), and reducing its characteristic wavelength. Interestingly enough, an increase of the electrical field increases the length of the jet too and, therefore, the jet diameter must decrease since the flow rate is kept constant.

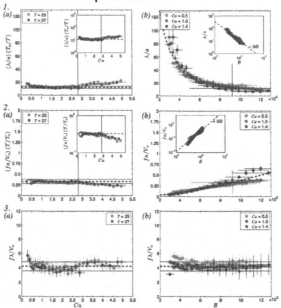

Figure 4: (1) Wavelength λ, (2) frequency f and (3) phase velocity $f\lambda$ of the whipping jets as a function of the capillary number (a) and the electrical Bond number (b). In (a) $\mathcal{B} = 920$ and $\mathcal{T} = 23$; in (b) $\mathcal{T} = 23$. White circles in (2a) correspond to results reported by Shin et al. (2001).

Whipping envelope

In this part, we analyze the behavior of the whipping envelope as a function of Ca and $\mathcal{B}$. Figure 2c shows an example of the whipping envelope. The amplitude A of the envelope corresponds to the maximum amplitude of the jet trajectory and the maximum angle of the droplet trajectories. We defined also the length h corresponding to the axial distance between the capillary end and the beginning of the whipping instability. We arbitrarily define the starting point of the whipping instability (the value of h) as the axial point at which the jet envelope reaches a threshold amplitude of $100 \ \mu m$, which corresponds to $6 - 7$ pixels in the images. Measurements were obtained by taking different values of the threshold; we have found that, in all cases, the relative

16

differences in the value of h were smaller than 15% for values of threshold amplitude within a range between 50 − 200 μm. The second length h_ω was defined as the axial distance from the point at which the jet bends and the whipping mode starts, $z = h$, until the jet breaks-up into droplets (see figure 1a). This length corresponds to the vertical (axial) length of the whipping envelope. These characteristic lengths have been obtained from data of the binary images and MATLAB algorithms. In figure 5, we have plotted the evolution of the whipping envelope as a function of Ca, column (a), and B, column (b). The amplitude A and the axial distance z are normalized with the needle diameter a. As shown in figure 5 (1a), two regimes may be identified: for $Ca \leq 2.8$, the amplitude A and length h_ω of the envelope increase with the capillary number while the envelope remains unchanged for $Ca > 2.8$. Note that the downstream behavior of the envelope changes also with the capillary number until the envelope reaches an asymptotic shape at $Ca = 2.8$ that remains unchanged for values of the capillary number larger than 2.8. The meridional shape of the different envelopes has been compared with that obtained in [6] using a PEO (Poly-Ethylene Oxide)-water mixture in air; the amplitude of the latter is represented by white crosses meridional line in figure 5 (1a). The dimensionless parameters of their experiment were $Ca = 0.4$, $\mathcal{T} = 2.4 \times 10^5$ and $B \sim O(10^1 - 10^2)$, which are larger than the minimum dimensionless values Ca_{min} and B_{min} for which the whipping instability develops in the case of glycerin-hexane. Note also that the dimensionless parameter $\mathcal{T}$ is greater than in the case of glycerin-hexane. The envelope obtained in [6] presents a different downstream behavior than the glycerin-hexane whipping envelopes. In the former case, the jet begins to bend at a smaller distance from the capillary, so that the envelope amplitude near the capillary is larger than in the latter case; also the amplitude of the envelope of the whipping jet is in their case two times larger than in ours. For $Ca > 2$, the envelope amplitude of the glycerine jet at $z > 40a$ becomes larger than that of the PEO-water mixture. Let us finally remark that the difference between the case reported in [6] and ours firstly lie in the fact that PEO-water mixture is a non-Newtonian fluid and secondly that the dimensionless time ratio $\mathcal{T}$ was 104 times larger in their case than in ours. Finally, the difference in the geometry of the electrical configuration (a plate-plate geometric configuration in theirs and a capillary-plate configuration in ours) may also play an important role in the envelope behavior. The shape of the whipping envelope depends on the electrical Bond number, 5 (1b). Essentially, the length h_ω of the whipping envelope decreases due to the increase of the length h with the electrical Bond number. For comparison, we have plotted a meridional curve of an envelope (white dashed line in figure 5 (1b)), which has been empirically obtained for the case of $B = 540$, see equation (4); the agreement with the experimental envelope is quite good. For a better comparison of the different experimental envelope curves, it proves convenient to plot the dimensionless amount $(A - r_j)/a$ versus $(z - h)/a$ with r_j the jet radius at $z = h$. As shown in figure 5 (2a), both the amplitude of the envelope and the length of the whipping increase with Ca whereas the whipping envelope is independent of B; therefore, the meridional curves of the envelope obtained at different values of B overlap. Also, the dependence on B of both h_ω and the maximum amplitude of the whipping envelope, $A_{max} = A(z = h_\omega)$, are rather small compared to their dependence on Ca. Note that, contrarily to the behavior of the shape of the envelope, the length h of the stable part of the jet depends on B, see figure 5 (1b). Observe also that the maximum angle θ that the droplet trajectories form with the jet axis (see figure 2c for the definition) depends on Ca but it is independent of B. For any combination of values of both Ca and B, the maximum angle of droplet trajectories was within the range of $\theta_{min} = 35°$ and $\theta_{max} = 70°$. The two extrema angles are plotted in figure 5 (2b).

17

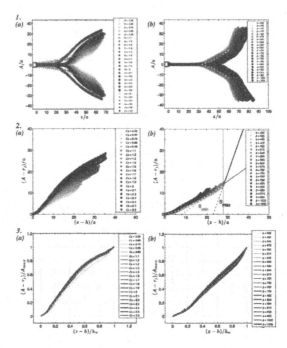

Figure 5: Amplitude of the envelope versus the axial distance as a function of (a) the capillary number and (b) the electrical Bond number. (1) Digital meridional images of the whipping envelope. (2) Meridional curves of the whipping envelope without the steady and the droplet region. (3) Normalized whipping envelopes.

Since whipping envelopes seem very similar to each other, we have normalized them with the maximum amplitude A_{max} while abscise $z - h$ has been also normalized using the whipping length h_ω. As a result of the rescaling, which are plotted in figure 5 (3a) the whipping envelopes collapse in a single curve. This self similar behavior of the envelopes has been also observed in experiments carried out as function of $\mathcal{B}$ at two other values of the capillary number, $Ca = 1.0$ and 1.4. Nonetheless, one should be aware that this self-similar behavior is observed only for a limited range of parameters: $0.55 \leq Ca \leq 2.8$, $400 \leq \mathcal{B} \leq 1055$ and $\mathcal{T} = 23 - 27$. We have calculated the mean curve of the normalized amplitude of the envelopes for each case: an example of the mean curves of the normalized envelopes is represented by dashed lines in figure 5 (3a&b); the two cases correspond respectively to the mean curve for a given electrical Bond number and several values of the capillary number, 5 (3a) and, that in 5 (3b), to the mean curve for a given capillary number and several values of the electrical Bond number. The mean curves allows comparing the mean amplitude of the envelopes for different values of Ca and $\mathcal{B}$.

18

In figure 6a, we have plotted for comparison the different mean curves of the envelopes in a log-log representation. In the range of $(z - h)/h_\omega$ between 10^{-2} and 10^{-1}, the mean envelope behaves linearly with the normalized axial distance and closely follows a law of the type $[(z - h)/h_\omega]^n$ for different electrical Bond and capillary numbers. Observe that this region $[(z - h)/h_\omega = O(10^{-2} - 10^{-1})$ is approximately of the order of 10 % at most of the total length of the envelope.

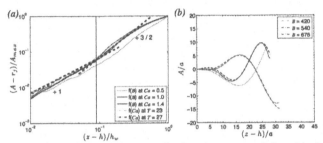

Figure 6: (a) Comparison of the normalized mean curves of the whipping envelopes from different experiments. (b) Comparison of the whipping jet images (straight line -) and the corresponding ones obtained from equation (4) (dashed line --) for different values of the electrical Bond number ($Ca = 0.5, \mathcal{T} = 23$).

Within the range of the abscissa $01 \le (z - h)/h_\omega \le 0.4$, which corresponds to the 30 % of the total length of the envelope, the different curves collapse in just one that follows a power law of the type $[(z - h)/h_\omega]^{3/2}$ provided that the capillary and the electrical Bond numbers are within the range of parameters $Ca \approx 0.5 - 2.8$ and $\mathcal{B} \approx 400 - 1055$. Let us, finally, remark that this result is quite different from the experimental observation reported in [6] for an electrified jet of a mixture of PEO (Poly-Ethylene Oxide) and water, for which they found a normalized whipping amplitude $A(z)/A_o$ following the exponential law $exp[wz/U_0]$. Using our experimental results, we have empirically modeled the amplitude of the whipping oscillation as,

$$A(z,t) = L\left(\frac{z-h}{c}\right)^\alpha cos\left[\left(\frac{2\pi(z-h)}{\lambda}\right)^\xi + 2\pi ft + \varphi\right], \tag{4}$$

where the exponents $\alpha = 3/2$ and $\xi \approx 1$ take into account the downstream evolution of the wavelength; c and L are characteristic lengths and φ is a phase amplitude. For our experiments, the mean value of c and L were $0.19\ mm$ and $0.78\ mm$, respectively. A comparison between several examples of the whipping jet images and the ones given by equation (4) are plotted in figure 6b. The calculated and experimental curves match quite well. Of course, the calculated envelope obtained by overlapping the calculated curves on one period ($1/f$) is also in good agreement with the experimental whipping envelopes (white dashed line in figure 5 (2a&b)). These results show the pertinence of expression (4) for modeling the whipping instability in the range of parameters $Ca \approx 0.5 - 2.8$ and $\mathcal{B} \approx 400 - 1055$.

CONCLUSIONS

We have presented new experimental measurements of the whipping instability exhibited by an electrified jet of glycerin travelling in a bath of hexane. With the help of a high-speed video camera, after proper image processing, we have measured different characteristic scales, frequency, wavelength and shapes of the whipping instability as a function of the two governing parameters: the electrical Bond number $\mathcal{B}$ and the capillary number Ca. We have observed the important effect of the electrical Bond number on the frequency and the wavelength, which prones the instability by increasing f and decreasing λ. Furthermore, the wavelength λ and frequency f appears to be independent of the capillary number and can be expressed as $\lambda = \lambda_o(\mathcal{B}/\mathcal{B}_o)^{-3/2} (\mathcal{T}/\mathcal{T}_o)$ and $f = f_o(\mathcal{B}/\mathcal{B}_o)^{3/2} (\mathcal{T}/\mathcal{T}_o)^{-1})$; where $\mathcal{B}_o = 880$, $\mathcal{T}_o = 23$, $\lambda_o = 12a$ and $f_o = V_o/3a$. In consequence, whatever the capillary number and electrical Bond number, the phase velocity of the whipping jet defined as the product of the dimensionless wavelength and frequency is constant and equal to $f\lambda = 4V_o$ with an error of $\pm 15\%$. That is, the phase velocity of the jet whipping depends only on the visco-capillary scale velocity V_o; its value for the glycerin-hexane couple is $8.8\ cm/s$. The envelope of the jet whipping has been also obtained from the experiments. The detected whipping envelope showed a self similar behavior after appropriate rescaling. The amplitudes of the different mean whipping envelopes normalized with the maximum amplitude evolved downstream towards a sole curve showing a $3/2$ power law of the normalize distance. The results illustrate the general behavior of these lateral instabilities and provide new insight that may be useful to understand the complex underlying dynamics. Finally, experiments with other couple of liquids should be carried out to confirm the different experimental behavior observed, in particular, the dependency of the frequency and wavelength or whipping jet behavior as function of the dimensionless parameters.

ACKNOWLEDGMENTS

The authors thank useful discussions with Prof. Higuera (UPM, Madrid, Spain). This project has been funded by the Spanish Ministry of Science and Technology under projects DPI2007-66659-C03-03 & 01, by The Junta de Andalucía under project P08-TEP-3997 and by Yflow Inc.

REFERENCES

1. A. Barrero and I. G. Loscertales, Annu. Rev. Fluid Mech. **39**, 89 (2007).
2. M. M. Hohman, M. Shin, G. Rutledge, and M. P. Brenner, Phys. Fluids, **13**, 2201 (2001).
3. D. H. Reneker, A. L. Yarin, H. Fong, and S. Koombhongse, J. Appl. Phys. **87**, 4531(2000).
4. A. L. Yarin, S. Koombhongse, and D. H. Reneker, J. Appl. Phys. **89**, 3018 (2001).
5. A. Barrero, J. M. López-Herrera, A. Boucard, I. G. Loscertales, and M. Márquez, J. Colloid Int. Sci. **272**, 104 (2004).
6. Y. M. Shin, M. M. Hohman, M. P. Brenner, and G. C. Rutledge, Polymer, **42**, 9955 (2001).
7. M. M. Hohman, M. Shin, G. Rutledge, and M. P. Brenner. Phys. Fluids, **13**, 2221 (2001).
8. A. G. Marín, I. G. Loscertales, M. Márquez, and A. Barrero, Phys. Rev. Lett. **98** (2007).
9. F. J. Higuera, J. Fluid Mech. **558**,143 (2006).
10. G. Taylor, Proc. R. Soc. A, **313**, 453 (1969).

Mater. Res. Soc. Symp. Proc. Vol. 1272 © 2010 Materials Research Society   1272-KK07-02

Electrospray of a very viscous liquid in a dielectric liquid bath

F. J. Higuera

E. T. S. Ingenieros Aeronáuticos, UPM, Pza. Cardenal Cisneros 3, 28040 Madrid, Spain

ABSTRACT

Numerical computations and order-of-magnitude estimates are used to analyze a jet of a very viscous liquid of finite electrical conductivity that is injected at a constant flow rate in an immiscible dielectric liquid under the action of an electric field. The conditions under which the injected liquid can form an elongated meniscus with a thin jet issuing from its apex (a cone-jet) are investigated by computing the flow, the electric field, and the transport of electric charge in the meniscus and a leading region of the jet. The boundaries of the domain of operation of the cone-jet mode are discussed. The current transfer region determining the electric current carried by the jet is analyzed taking into account the viscous drag of the dielectric liquid surrounding the jet. Conditions under which the electric current/flow rate characteristic follows a square root law or departs from it are discussed.

INTRODUCTION

Electrostatic dispersion is a technique used to generate emulsions of electrically conducting liquids in baths of immiscible dielectric liquids. It leads to drops with narrow distributions of sizes which can be controlled in the range from a few tens of nanometers to hundreds of micrometers, and therefore has potential applications to the synthesis of nanoparticles, the encapsulation of antibacterial or antifungal agents, and the manufacturing of drug-laden particles for targeted delivery, among others. The technique originated in the work of Barrero and coworkers [1, 2], who showed that the cone-jet mode of an electrospray, which is often used to atomize conducting liquids in air or vacuum [3-8], can also be used in a bath of a dielectric liquid. These authors investigated fundamental properties of the cone-jet mode in a liquid bath, such as the current/flow rate and current/voltage characteristics, the scaling laws for the droplet size, and the enhanced whipping instability of the jets of these electrosprays compared with their counterparts in air. Gundabala and Fernández-Nieves [9] have carried out work along these lines for electrosprays embedded in microfluidic channels, and Alexander [10] investigated the pulsating modes of an electrospray in a liquid bath.

In this paper, numerical computations and order-of-magnitude estimates are used to fur-

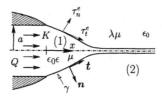

Figure 1: Sketch of the meniscus and the jet near the end of the capillary tube.

ther analyze the problem and propose scaling laws. Attention is focused on the stationary current transfer region of the electrospray's jet where convection of the electric charge accumulated at the interface between the two liquids begins to account for a fraction of the electric current of the spray. A square root current/flow rate characteristic is obtained in a certain range of flow rates, but the viscous drag due to the liquid bath may limit the length of the straight stationary jet and cause the electric current to fall below this square root law for large values of the flow rate.

FORMULATION

Figure 1 is a sketch of the system. The liquid to be dispersed (liquid 1) has viscosity μ, electrical conductivity μ and permittivity $\epsilon_0\epsilon$. It is injected at a constant flow rate Q through a metallic capillary tube of radius a into a bath of an immiscible dielectric liquid (liquid 2) of viscosity $\lambda\mu$ and permittivity ϵ_0 which is at rest far from the tube. Liquid 1 forms a meniscus at the end of the tube, which is subject to an electric field due to a high voltage applied between the capillary and a far electrode immersed in the bath. The electric field induces a conduction current in liquid 1 which accumulates electric charge at the interface between the two liquids. The field also causes electric stresses at the interface which must be balanced by pressure, viscous and surface tension stresses. The electric stresses elongate the meniscus in the direction of the field and, in certain conditions, lead to the formation of a thin jet that emerges from the tip of the meniscus and breaks into drops at some distance downstream. The effect of the inertia of the liquids is left out of the following analysis, which is a simplification appropriate for very viscous liquids in the meniscus and in an initial region of the jet; see, e.g., Ref. [11]. The flow in these regions is assumed to be stationary and axisymmetric.

In these condition, the governing equations are

$$\nabla^2\varphi_i = 0 \qquad (1)$$

$$\nabla \cdot \boldsymbol{v}_i = 0, \quad 0 = -\nabla p_i + \mu_i \nabla^2 \boldsymbol{v}_i, \qquad (2)$$

22

where the subscript i denotes the variables in each liquid; $i = 1, 2$ with $\mu_1 = \mu$ and $\mu_2 = \lambda\mu$. The boundary conditions at the interface between the two liquids are

$$\epsilon_0 \left(E_{2n} - \epsilon E_{1n} \right) = \sigma, \quad E_{1t} = E_{2t} \tag{3}$$

$$\mathbf{v}_1 \cdot \boldsymbol{\nabla}\sigma = K E_{1n} + \sigma \mathbf{n} \cdot \boldsymbol{\nabla}\mathbf{v}_1 \cdot \mathbf{n} \tag{4}$$

$$\mathbf{v}_1 \cdot \boldsymbol{\nabla} f = 0, \quad \mathbf{v}_1 = \mathbf{v}_2 \tag{5}$$

$$p_2 - p_1 + \mathbf{n} \cdot (\tau_1' - \tau_2') \cdot \mathbf{n} + \gamma \boldsymbol{\nabla} \cdot \mathbf{n} = \tau_n^e \tag{6}$$

$$\mathbf{t} \cdot (\tau_1' - \tau_2') \cdot \mathbf{n} = \tau_t^e. \tag{7}$$

Equations (6) and (7) are balances of stresses normal and tangent to the interface. Here p_1 and p_2 are the pressures of the liquids, τ_1' and τ_2' are the viscous stress tensors, and γ is the interfacial tension The electric stresses normal and tangent to the interface, in the right-hand sides of these equations, are (Landau and Lifshitz [12] and Saville [13])

$$\tau_n^e = \frac{\epsilon_0}{2} \left(E_{2n}^2 - \epsilon E_{1n}^2 \right) + \frac{\epsilon_0}{2}(\epsilon - 1)E_{1t}^2 \quad \text{and} \quad \tau_t^e = \sigma E_{1t} \tag{8}$$

in terms of the density of free surface charge σ and the normal and tangent components of the electric field at the two sides of the interface ($E_{in} = \mathbf{E}_i \cdot \mathbf{n}$ and $E_{it} = \mathbf{E}_i \cdot \mathbf{t}$, where $\mathbf{n}$ and $\mathbf{t}$ are unit vectors normal and tangent to the interface; see Fig. 1).

The electric field in liquid i is $\mathbf{E}_i = -\boldsymbol{\nabla}\varphi_i$, where the electric potentials φ_i satisfy the Laplace's equations (1), to be solved with the electrostatic boundary conditions (3) at the interface, the condition that the surface of the tube is an equipotential, and that the electric potential away from the tube tends to the potential of a needle, $\varphi_\infty = V^* \ln \left\{ \left[(x^2 + r^2)^{1/2} + x \right] / a \right\}$, where V^* is proportional to the voltage applied between the electrodes and x and r are distances along the axis of the tube and normal to the axis.

The density of surface charge obeys the transport equation (4), which expresses the condition that the charge of a material element of the interface increases at the rate at which conduction in the inner liquid (liquid 1) accumulates charge at the interface.

The velocity and pressure of each liquid are given by the mass and momentum conservation equations (2), to be solved with the boundary conditions (5)–(7) at the interface and the conditions that the velocity of the outer liquid (liquid 2) is zero at the outer surface of the tube and at infinity and that the velocity of the inner liquid is $Q/(\pi a^2)$ at the outlet of the tube ($x = 0$). Conditions (5) mean that the interface, which is sought in the form $r = r_s(x)$ with $r_s(0) = a$, is a material surface.

The electrical and mechanical problems are coupled through the electric stresses at the interface.

The problem can be written in dimensionless variables using the radius of the tube a, the viscous-capillary velocity $v_c = \gamma/\mu$, and the field $E_c = (\gamma/\epsilon_0 a)^{1/2}$ as scales of length, velocity and electric field. The solution depends on the five dimensionless parameters

$$Ca = \frac{\mu Q}{\gamma a^2}, \quad V = \frac{\epsilon_0^{1/2} V^*}{\gamma^{1/2} a^{1/2}}, \quad \epsilon, \quad \Lambda = \frac{\mu K a}{\epsilon_0 \gamma}, \quad \lambda. \quad (9)$$

RESULTS AND DISCUSSION

Numerical results

Figure 2 shows different elements of the solution for $V = 1.8$, $\epsilon = 20$, $\Lambda = 10^3$, $\lambda = 0.03$, and three values of the dimensionless flow rate Ca, as functions of the distance x to the tube. Figure 1(a) shows longitudinal sections of the meniscus and the jet. As can be seen, the size of the meniscus and the radius of the jet increase with the flow rate.

Figure 2(b) shows the contributions of conduction in the inner liquid ($I_b = 2\pi K \int_0^{r_s} E_{1x} r \, dr$, where E_{1x} is the axial field in the inner liquid; dashed curves) and convection of the surface charge ($I_s = 2\pi \sigma v r_s$; solid curves) to the electric current. The sum $I = I_b + I_s$ is a constant, equal to the current carried by the jet. The conduction current I_b dominates in the meniscus and the convection current I_s dominates in the jet where, however, the conduction current does not seem to go to zero. The cross-over point where the two contributions to the current are equal to each other is well into the jet and shifts streamwise when the flow rate increases.

Figure 2(c) shows the electric stress tangent to the surface (dashed curves) and the difference between the normal electric stress and the surface tension stress (solid curves; here $\mathcal{C}$ is twice the mean curvature of the interface). The tangent electric stress drives the flow in the jet. However, this force is small in the meniscus, where the interface is nearly equipotential and the tangent electric field is small, and rises to a maximum in the current transfer region around the cross-over point where the two contributions to the current are of the same order.

The momentum equation for the inner liquid in (2) ($i = 1$) can be integrated across the jet, where the flow is quasi-unidirectional, to give the balance of forces (see, e.g., Gañán-Calvo et al. [14] and Feng [15])

$$\frac{\partial}{\partial x}\left(3\pi r_s^2 \mu \frac{\partial v}{\partial x}\right) + \pi r_s^2 \frac{\partial}{\partial x}\left(\tau_n^e - \frac{\gamma}{r_s} + \tau_n^o\right) + 2\pi r_s \left(\tau_t^e + \tau_t^o\right) = 0, \quad (10)$$

where the terms on the left-hand side are the axial forces per unit length of the jet due to the effective axial viscous stress in the inner liquid; to the axial gradient of the pressure variation induced by the normal electric stress, the surface tension stress and the stress of the outer liquid normal to the interface ($\tau_n^o = -p_2 + \boldsymbol{n} \cdot \tau_2' \cdot \boldsymbol{n}$); and the forces due to the electric shear stress τ_t^e and the viscous shear stress of the outer liquid ($\tau_t^o = \boldsymbol{t} \cdot \tau_2' \cdot \boldsymbol{n}$).

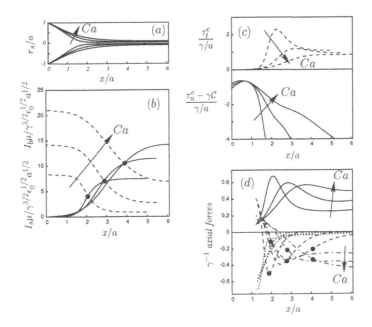

Figure 2: (a) Longitudinal sections of the interface. (b) Convection (solid) and conduction (dashed) contributions to the electric current nondimensionalized with $\epsilon_0 E_c v_c a$ [see definitions above Eq. (9)]. (c) Stresses normal ($\tau_n^e - \gamma\mathcal{C}$, solid) and tangent ($\tau_t^e$, dashed) to the interface nondimensionalized with γ/a. (d) Axial forces acting on the jet per unit streamwise length nondimensionalized with γ. Electric force $2\pi r_s \tau_t^e$ in the last term of (10) (solid); force due to the axial viscous stress in the inner liquid, $\partial(3\pi r_s^2 \mu \partial v/\partial x)/\partial x$ in (10) (dashed); force due to the normal electric stress and the surface tension, $\pi r_s^2 \partial(\tau_n^e - \gamma/r_s)/\partial x$ in (10) (dotted); and force due to the viscous shear stress of the outer liquid, $2\pi r_s \tau_t^o$ in the last term of (10) (chain). Values of the parameters are $Ca = 0.0256$, 0.0858 and 0.1977 (increasing as indicated by the arrows), $V = 1.8$, $\epsilon = 20$, $\Lambda = 10^3$ and $\lambda = 0.03$. The black circles in (b) and (d) mark the crossover point at which the two contributions to the current are equal to each other.

25

These forces are evaluated from the full numerical solution and shown in Fig. 2(d). In the current transfer region, the force due to the electric shear stress (solid curves) is balanced mainly by the forces due to the axial viscous stress of the inner liquid (dashed curves) and the viscous shear stress of the outer liquid (chain curves). The values of these two forces at the current cross-over point are marked by black circles. As can be seen, the force due to the viscosity of the inner liquid dominates at small flow rates and the drag of the outer liquid becomes more important when the flow rate increases. The first of these forces tends to zero far downstream, leaving a balance of electrical and viscous shear stresses in the far jet.

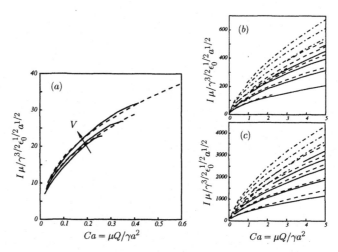

Figure 3: (a) Current/flow rate characteristic for $\lambda = 0.01$, $V = 1.6$ and 1.8 (dashed curves), and for $\lambda = 0.03$, $V = 1.6$, 1.8 and 2 (solid curves). Values of other parameters are $\epsilon = 20$ and $\Lambda = 10^3$. (b and c) Current/flow rate characteristic for two parallel plate electrodes. Here $\lambda = 0$ (solid), 0.01 (dashed) and 0.1 (chain), $E_\infty = 1, 2, 3, 4, 5$, increasing from bottom to top, $\epsilon = 20$, and $\Lambda = 10^3$ (b) and 2×10^4 (c).

The electric current carried by the jet is shown in Fig. 3(a) as a function of the dimensionless flow rate Ca for different values of the other dimensionless parameters. The current increases nearly as the square root of the flow rate in a certain range of this variable. The current also increases with the applied voltage V and with the ratio of outer-to-inner liquid viscosity λ.

There is no stationary solution below a certain minimum flow rate, near which the meniscus is conical and the jet is very thin. There is also a maximum flow rate near which the current falls below the square root law and seems to level to a constant value before the computations cease to converge to a stationary solution. The departure from the square root law is more pronounced and occurs earlier when the viscosity of the outer liquid increases. The volume of the meniscus and the radius of the jet increase with the flow rate, and the meniscus does not resemble a Taylor cone at the maximum flow rate.

Figures 3(b,c) show the electric current as a function of the flow rate for a parallel plate electrode configuration, for which the electric field tends to a constant (E_∞ in dimensionless variables) far from the meniscus. The results for this case are similar to those of Fig. 3(a) for a needle-plate electrode configuration, but the current increases with a different power of the flow rate, which is about 2/3.

Order-of-magnitude estimates

The following three conditions are satisfied in the current transfer region of the jet.

- The conduction and convection contributions to the current are of the same order. This condition, which defines the current transfer region, can be written in the form

$$KE_T r_s^2 \sim \sigma v r_s \quad \text{with} \quad E_T \sim \left(\frac{\gamma}{\epsilon_0 x}\right)^{1/2} \quad \text{and} \quad v \sim \frac{Q}{r_s^2}. \tag{11}$$

Here E_T is the electric field acting on the jet, which is the Taylor field due to the conical meniscus. The velocity of the liquid can be written in terms of the flow rate and the local radius of the jet.

- The axial electric field induced by the charge at the surface of the jet, which acts as a line distribution of charge, is of the order of the Taylor field E_T and partially balances it at the surface. In orders of magnitude and leaving out a logarithmic factor [16], this condition reads

$$E_{2n}\frac{r_s}{x} \sim E_T \quad \text{with} \quad \epsilon_0 E_{2n} \sim \sigma. \tag{12}$$

This screening condition is a consequence of the conductivity of the inner liquid, which tends to accumulate electric charge at the interface until this charge screens the inner liquid from the outer field and conduction ceases. However, the screening cannot be complete over the whole jet when the liquid is in motion, because the flow removes surface charge from any given region of the interface and this charge must be continuously replaced by new charge carried to the interface by conduction in the inner liquid. This, in turn, requires an electric field in the inner liquid, and this field increases with the length of the region to be screened. Since the inner field cannot be larger than the outer field, the condition that the two fields be of the same order determines the length of the region that can be partially screened, which coincides with the current transfer region where the outer field E_T first enters the inner liquid.

27

- The third condition to be satisfied is a balance of forces. The force driving the flow in the current transfer region, due to the electric shear stress at the interface, is of order $r_s \sigma E_T$. In the absence of inertia, this force must be balanced by (a) the force due to the axial viscous stress in the inner liquid (first term of (10), which is of order $r_s^2 \mu v / x^2$), or (b) the force due to the viscous shear stress of the outer liquid (last term of (10), which is of order $r_s \lambda \mu v / r_s$, where it is assumed that the thickness of layer of outer liquid set in motion by the electric force is of the order of the radius of the jet or larger than it; see, e.g., Ref. [11] for details).

The relevant balance of forces, together with conditions (11) and (12), determines the characteristic length of the current transfer region and the characteristic radius of the jet in this region as

$$x \sim \frac{\mu^{1/2} Q^{1/2}}{\gamma^{1/2}}, \quad r_s \sim \frac{\epsilon_0^{1/4} \mu^{1/8} Q^{3/8}}{K^{1/4} \gamma^{1/8}} \tag{13}$$

in case (a) and

$$x \sim \frac{\lambda^2 \mu^2 K Q}{\epsilon_0 \gamma^2}, \quad r_s \sim \frac{\lambda^{1/2} \mu^{1/2} Q^{1/2}}{\gamma^{1/2}} \tag{14}$$

in case (b).

The characteristic current carried by the jet follows the square root law

$$I \sim (\gamma K Q)^{1/2} \tag{15}$$

in the two cases.

Which of the two possibilities (13) or (14) is realized for a given set of values of the parameters can be determined by comparing the estimates of the forces in (10). Thus,

$$\frac{2\pi r_s \tau_t^o}{\partial (3\pi r_s^2 \mu \partial v / \partial x) / \partial x} \sim \Pi = \frac{\lambda \mu^{3/4} K^{1/2} Q^{1/4}}{\epsilon_0^{1/2} \gamma^{3/4}} \tag{16}$$

when (13) is used. Therefore the drag of the outer liquid is the dominant force to be balanced by the electric shear in the current transfer region if Π is large compared to unity, and this drag is negligible in the current transfer region if Π is small compared to unity.

The condition used in (12), that the density of surface charge in the current transfer region is of the order of the equilibrium charge density that would electrically screen the inner liquid, requires that the time available for conduction to charge the interface, which is of the order of the residence time $t_r = x/v$, be large compared to the electric relaxation time of the inner liquid $t_e = \epsilon_0 \epsilon / K$. The ratio t_r / t_e is an increasing function of the flow rate in the two possible cases. The condition that this ratio be of order unity determines

28

the order of the flow rate at which charge relaxation effects come into play in the current transfer region as

$$Q_{min}^{(a)} \sim \frac{\epsilon^4 \epsilon_0^2 \gamma^3}{\mu^3 K^2} \quad \text{or} \quad Q_{min}^{(b)} \sim \frac{\epsilon \epsilon_0^2 \gamma^3}{\lambda^3 \mu^3 K^2}. \tag{17}$$

These estimates give also a lower bound of the range of flow rates where the square root law for the current (15) can be realized. Which of the two estimates (17) is relevant depends on the value of the product $\lambda \epsilon$. If this product is small, then the drag of the outer liquid becomes small compared to the force due to the axial viscous stress in the inner liquid before charge relaxation effects appear, which happens when the flow rate decreases to values of order $Q_{min}^{(a)}$. On the other hand, the drag of the outer liquid dominates and charge relaxation effects appear for flow rates of order $Q_{min}^{(b)}$ if the product $\lambda \epsilon$ is large.

Convection of the surface charge makes an important contribution to the total electric current downstream of the current transfer region. This determines the order of the density of surface charge in the far jet as $\sigma \sim I r_s / Q$, and the electric shear stress as $\tau_t^e = \sigma E_\infty(x)$, where $E_\infty(x)$ is the electric field acting on the jet in this far region. In addition, the equilibrium of the driving electrical force and the viscous force of the inner liquid or the drag of the outer liquid determines the evolution of the radius of the jet as

$$\frac{r_s^2 \mu v}{x^2} \sim r_s \tau_t^e \Rightarrow r_s \sim \left(\frac{\mu Q^2}{I x^2 E_\infty} \right)^{1/2} \tag{18}$$

or

$$\frac{\lambda \mu v}{r_s} \sim \tau_t^e \Rightarrow r_s \sim \left(\frac{\lambda^2 \mu^2 Q^2}{I E_\infty} \right)^{1/4}. \tag{19}$$

The second possibility, corresponding to the equilibrium of electric shear stress and viscous shear stress of the outer liquid, should be always realized sufficiently far downstream for any $\lambda > 0$. However, the jet may break into drops or undergo other instability before reaching the final state (19). If this state is attained and the field E_∞ acting on the far jet is nearly uniform, which is probably the case in the experiments of Riboux et al. [2] carried out with a needle-plate electrode configuration, then the radius of the jet and the conduction current are predicted to reach constant values independent of streamwise distance. This result agrees with the experimental observations in Ref. [2].

The first result (13) predicts that the length of the current transfer region increases linearly with the flow rate in the case when the drag of the outer liquid is large compared to the force due to the axial viscous stress of the inner liquid in the current transfer region. The length x becomes then of the order of the radius of the capillary a when the flow rate becomes of the order of

$$Q_{max} = \frac{\epsilon_0 \gamma^2 a}{\lambda^2 \mu^2 K}. \tag{20}$$

The estimates leading to (13) should be modified for $Q \gg Q_{max}$ because the transfer of current to the surface of the jet would occur in a region where the electric field acting on the jet is not the field of a Taylor cone but decreases as the inverse of the distance to the tube. This changes the electric force on the jet, and the balance of this force and the drag of the outer liquid gives a radius of the jet than increases with streamwise distance upstream of the current transfer region, where the electric shear stress is still small and the force driving the flow is due to the gradient of the depression induced in the inner liquid by the electric stress normal to its surface [second term of (10)]. But this result is meaningless because the latter force points toward the meniscus when r_s increases with x. This negative result means that the straight stationary jet cannot extend to distances downstream of the tube much larger than its radius a. The jet is expected to break up or undergo whipping almost immediately upon being injected when Q is of the order of Q_{max} or larger. The current that can be transferred to the surface in the limited region where the jet is straight and stationary can be easily estimated and turns out to be independent of the flow rate, which is in qualitative agreement with experiments [2, 9].

The numerical results show that the ratio of the force due to the electrically induced depression to the drag of the outer liquid decreases at the current cross-over point when the flow rate increases. This means that the equilibrium of forces in the current transfer region depends more and more on the electric shear stress. Apparently this equilibrium ceases to be possible when the rise of the electric shear stress is postponed, by increasing the flow rate, beyond the region where the electric field is that of a Taylor cone.

SUMMARY

The flow, the electric field and the surface charge of a stationary electrified jet of a very viscous liquid injected into a quiescent immiscible dielectric liquid have been computed numerically.

Qualitative estimates have been worked out for the current transfer region of the jet under different conditions of the flow.

A square root current/flow rate characteristic is found in a certain range of flow rates.

Charge relaxation effects come into play in the current transfer region when the flow rate approaches a certain minimum which depends on the properties of the liquids.

A maximum flow rate is found above which no long stationary jet can exist.

ACKNOWLEDGMENTS

This work was supported by the Spanish Ministerio de Educación y Ciencia through

Project No. DPI2007-66659-C03-02.

References

[1] A. Barrero, J. M. López-Herrera, A. Boucard, I. G. Loscertales and M. Márquez, *J. Colloid Interface Sci.* **272**, 104 (2004).

[2] G. Riboux, A. G. Marín, I. G. Loscertales and A. Barrero, *J. Fluid Mech.* Submitted (2010). Also Procs. 1st European Conf. on Microfluidics, Bologna, December 10–12 (2008).

[3] J. Zeleny, *Proc. Camb. Phil. Soc.* **18**, 1 (1015).

[4] G. I. Taylor, *Proc. R. Soc. Lond. A* **280**, 383 (1964).

[5] D. P. H. Smith, *IEEE Trans. Ind. Appl.* **IA22**, 527 (1986).

[6] M. Cloupeau and B. Prunet-Foch, *J. Electrost.* **22**, 135 (1989).

[7] J. Fernández de la Mora and I. G. Loscertales, *J. Fluid Mech.* **260**, 155 (1994).

[8] J. Fernández de la Mora, *Ann. Rev. Fluid Mech.* **39**, 217 (2007).

[9] V. R. Gundabala and A. Fernández-Nieves, *Phys. Rev. Lett.* Submitted (2010).

[10] M. S. Alexander, *Appl. Phys. Lett.* **92**, 144102 (2008).

[11] F. J. Higuera, *J. Fluid Mech.* **648**, 35 (2010).

[12] L. D. Landau and E. M. Lifshitz, *Electrodynamics of Continuous Media* (Pergamon, Oxford, 1960).

[13] D. A. Saville, *Ann. Rev. Fluid Mech.* **29**, 27 (1997).

[14] A. M. Gañán-Calvo, J. Dávila and A. Barrero, *J. Aerosol Sci.* **28**, 249 (1997).

[15] J. J. Feng, *Phys. Fluids* **14**, 3912 (2002).

[16] E. J. Hinch, *Perturbation Methods* (Cambridge University Press, Cambridge, 1991).

Mater. Res. Soc. Symp. Proc. Vol. 1272 © 2010 Materials Research Society   1272-KK08-02

Droplet Based Microfluidics for Synthesis of Mesoporous Silica Microspheres

Nick J. Carroll,[1] Svitlana Pylypenko,[2] Amber Ortiz,[1] Bryan T. Yonemoto,[1] Ciana Lopez,[1] Plamen B. Atanassov,[2] David A. Weitz,[3] and Dimiter N. Petsev[1*]

[1]Department of Chemical and Nuclear Engineering and Center for Biomedical Engineering, University of New Mexico, Albuquerque, New Mexico 87131

[2] Center for Emerging Energy Technologies, University of New Mexico, Albuquerque, New Mexico 87131

[3]School of Engineering and Applied Sciences, Department of Physics, Harvard University, Cambridge, Massachusetts 02138

ABSTRACT

Herein we present methods for synthesizing monodisperse mesoporous silica particles and silica particles with bimodal porosity by templating with surfactant micelle and microemulsion phases. The fabrication of monodisperse mesoporous silica particles is based on the formation of well-defined equally sized emulsion droplets using a microfluidic approach. The droplets contain the silica precursor/surfactant solution and are suspended in hexadecane as the continuous oil phase. The solvent is then expelled from the droplets, leading to concentration and micellization of the surfactant. At the same time, the silica solidifies around the surfactant structures, forming equally sized mesoporous particles. We show that hierarchically bimodal porous structures can be obtained by templating silica microparticles with a specially designed surfactant micelle/microemulsion mixture. Oil, water, and surfactant liquid mixtures exhibit very complex phase behavior. Depending on the conditions, such mixtures give rise to highly organized structures. A proper selection of the type and concentration of surfactants determines the structuring at the nanoscale level. Tuning the phase state by adjusting the surfactant composition and concentration allows for the controlled design of a system where microemulsion droplets coexist with smaller surfactant micellar structures. The microemulsion droplet and micellar dimensions determine the two types of pore sizes.

INTRODUCTION

Emulsification of a polymer precursor followed by execution of the polymer chemistry within emulsion droplet reactors provides a facile and versatile method for producing microparticles. Not surprisingly, if a liquid-to-solid chemical reaction proceeds to completion within these drops, the resultant solid particles will possess the shape of the droplets.[1, 2] Microfluidic flow-focusing devices (MFFDs) provide a straightforward and robust approach to the formation of highly monodisperse emulsion drops.[3] It has been demonstrated that microfluidic-generated drops can function as both morphological templates and chemical reactors for the synthesis of monodisperse polymer [4-6] and biopolymer [7] particles.

An appealing feature to engineer into emulsion-polymerized particles is porosity. Particles with well-defined pore morphology are essential for many areas of modern technology. Potential applications include catalysis [8, 9] and electrocatalysis,[10] chromatography,[11] and drug delivery.[12] Precise control over the pore size and shape is crucial for the successful performance of the particles. It allows for optimization of fluid transport in a catalyst, determines the molecular release of solute by a drug delivery vehicle, or defines the size selectivity in chromatography. Templating of oxide materials with surfactant micelles is a powerful method to obtain mesoporous structures with controlled morphology.[13] Oxide (e.g., silica) precursor solution (sol) is mixed with a templating surfactant, and evaporation of the solvent leads to an increase in the surfactant concentration. The surfactant forms supramolecular structures according to the solution phase diagram. This is known as evaporative induced self-assembly (EISA) and has been used to obtain bulk porous materials or microparticles using high temperature aerosol methods.[14-16]

Alternatively, mesoporous particle synthesis via EISA can be performed in water in oil emulsion droplets under milder temperature stresses. Recently, Andersson et al.[17] demonstrated the synthesis of spherical mesoporous silica particles using an approach that combines previously established emulsion-based precipitation methods [18, 19] with the EISA method. This synthesis route, referred to as the emulsion and solvent evaporation method (ESE), produced well-ordered 2D hexagonal mesoporous silica microspheres. The advantages of this method are control of synthesis parameters such as emulsion droplet size, temperature, evaporation speeds, humidity, and the composition of the surfactant solution. In comparison to aerosol methods, the relatively slower evaporation rate of the solvent from the emulsion droplets allows a high-degree of homogeneity of the components in the liquid crystalline phase prior to fossilization of the structures by silica condensation. This is perhaps the most important distinction of the emulsion EISA method from aerosol-based EISA methods.

Surfactant self-assembly provides a powerful method for synthesizing mesoporous materials. However, these materials are limited to micelle-templated pores with diameters of a few nanometers, and mesoporous microparticles synthesized by aerosol and ESE methods enclose internal structures rendered inaccessible at the surface due to inherent formation of a solid material layer at the surface.[16, 17] To address the requirements of emerging technologies, the next generation of porous oxide materials must be highly structured and functionalized. Hierarchically porous structures offer advantages in design of materials where catalytic activity is to be utilized in immediate conjunction with transport of reactants. Templating approaches for hierarchical material fabrication are attractive, as they can be combined with other methods such as impregnation or precipitation to yield structures with controlled porosity, surface chemistry, and hydrophilicity or hydrophobicity. Interfacial phenomena such as spontaneous formation of complex microemulsion phases [20, 21] present exceptional and generally less-explored avenues for particle nanostructure templating.

This proceedings paper describes the formation of mesoporous silica particles via two different synthesis routes. The first is based on MFFD emulsification of an aqueous-based sol with subsequent processing utilizing the ESE method to produce monodisperse silica particles with a single pore size distribution with dimensions of a few nanometers.[22] The second method facilitates design of mesoporous microparticles with biporous internal structure with one set of pores with dimensions in the tens of nanometers and a subset of smaller pores

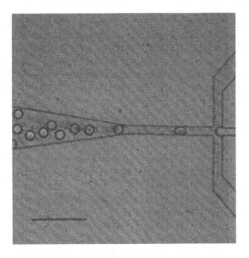

Figure 1. Optical microscopy image of droplets of silica precursor solution emulsified in a T-shaped microfluidic device in hexadecane. The channel dimensions of the orifice are 25 μm (width) by 30 μm (height). The scale bar is 100 μm.

with dimensions of a few nanometers.[23] The two types of pores are due to the coexistence of oil microemulsion droplets with smaller ionic micelles in aqueous drops of a few micrometers in diameter. This has been accomplished by tuning the phase state of the oil/water/surfactant mixture to ensure the presence of the different species.

Experiment

For synthesis of monodisperse, surfactant-templated mesoporous silica particles, the silica precursor solution was prepared by hydrolyzing 5.2 g of tetraethylorthosilicate (TEOS, Purum >98%) in 3 g of ethanol (99.7%) and 2.7 g of 0.01 N hydrochloric acid (pH 2) under vigorous stirring at room temperature for 30 min. Next, 1.4 g of the amphiphilic triblock copolymer templating molecule (Pluronic, BASF, P104) was dissolved in 5.43 g of DI water and subsequently mixed with the hydrolyzed TEOS solution to complete the preparation of the aqueous-based sol. We designed this particular recipe to allow the use of Pluronic surfactant as a templating reagent in the presence of a much lower concentration of ethanol than used by other authors[17]. Emulsification of the aqueous siliceous precursor was achieved by supplying the sol-dispersed phase and organic oil continuous phase to the microfluidic device using two digitally controlled Harvard Pico Plus syringe pumps. The droplet (and therefore particle) production was approximately 100/s. The continuous phase was prepared by dissolving ABIL EM 90 (Degussa) surfactant in hexadecane (3 wt %), which served as an emulsion stabilizer. The volumetric flow rate for the dispersed sol was optimized to 0.5 μL/min, with a flow rate of 3.5 μL/min for the continuous oil phase. The SU-8 photoresist-templated poly(dimethylsiloxane) (PDMS) microfluidic device was fabricated using a well-established softlithography method. The microfluidic device used in this study is shown in Figure 1. The MFFD-produced droplets were transferred to a 50mL round bottomed flask and heated to 80 °C under a reduced pressure of 70 mTorr for 2 h. The flask was pretreated with RAIN-X solution to make it hydrophobic. This was necessary to prevent the droplets from sticking to the flask bottom. To prevent droplet flocculation and coalescence before the sol-gel transition was complete, the emulsion was subjected to constant stirring at 200 rpm. This stirring was sufficient to keep the droplets suspended and separated, and did not lead to shear deformation. The particles were then collected and centrifuged, followed by calcination in air at 500 °C for 5 h to remove the templating surfactant.

For synthesis of microemulsion-templated silica particles with bimodal porosity, the silica precursor solution was prepared by dissolving 1.82 g of cetyltrimethylammonium bromide (CTAB) in 20 g of DI water under vigorous stirring at 40°C until the solution was clear. Next, 5.2 g of TEOS (Purum>98%) and 0.57 g of 1 N hydrochloric acid were added to the mixture under vigorous stirring at room temperature for 30 min to hydrolyze and dissolve the TEOS monomer. The measured acidity of the hydrolyzed sol showed pH ≈ 2. The oil phase was prepared by dissolving a modified polyetherpolysiloxane/dimethicone copolyol surfactant with the trade name ABIL EM 90 (Degussa) in hexadecane (3 wt %). The aqueous siliceous precursor solution was then added to the oil phase and subsequently emulsified by brisk shaking of the vial. The emulsion was transferred to a 1000 mL round-bottom flask and heated to 80°C under a reduced pressure of 70mTorr for 3 h. The particles were collected and centrifuged, and the supernatant oil removed, followed by calcination in air at 500°C for 5 h to remove the templating surfactant.

Dynamic light scattering (DLS) studies were conducted on a Nanotrac NPA250 dynamic light scattering instrument from Microtrac Inc. The measurements were separately performed in the two macroscopic phases (aqueous solution of the CTAB and silica precursor that has been in contact with the oil containing the ABILEM90). After waiting for 48 h, the oil phase spontaneously dispersed into the water in the absence of any additional stirring. The X-ray powder diffraction (XRD) patterns were obtained on a Scintag diffractometer (Cu KR radiation). Transmission electron microscopy (TEM) was conducted on JEOL 2010 and 2010F instruments, and scanning electron microscopy (SEM) was done on a Hitachi S-800 instrument. Nitrogen (77.4) adsorption/desorption measurements were performed on a Quantochrome Autosorb-I-MP instrument. Prior to analysis, the sample was outgassed overnight at 120°C.The adsorption data were analyzed using an NLDFT approach and cylindrical pore model.

DISCUSSION

Figure 2 shows silica particles obtained from TEOS precursor droplets formed in hexadecane oil. The hexadecane allows for the solvent (DI water and ethanol) to be expelled from the droplets, which leads to polymerization of the templated silica. Figure 2A shows particles that were obtained from a shaken bulk emulsion. The particles have a well-defined spherical shape but are very polydisperse. It is not possible to improve on this because a shaken emulsion produces polydisperse droplets, which later result in a wide particle size distribution. Using a microfluidic device (Figure 1), we were able to form monodisperse droplets, which is the necessary condition for obtaining well-defined monodisperse particles. The size of the droplets depends on the dimensions of the microchannel, the flow rates in the central and side channels, the viscosity of the fluids (water/ethanol and oil), and the surfactant. Hence, a single device can

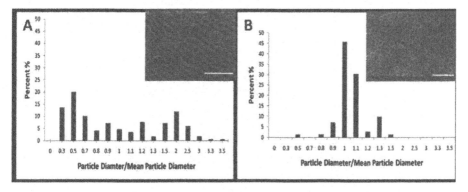

Figure 2. (A) Size distribution and SEM image of polydisperse silica microspheres templated from a shaken bulk emulsion. The scale bar is 20 μm. (B) Size distribution and SEM image of silica microspheres templated by monodisperse microfluidic device-generated droplets. The scale bar is 100 μm. Reproduced from Ref. 22.

produce monodisperse droplets of different sizes by varying the relative magnitude of the viscous and interfacial forces that are involved. We were able to vary the droplet size by 1 to 3 times the channel width. Hence, to cover a wider range of sizes one may need more than one microfluidic device or may have several channels, each with different sizes, fabricated on the same device. The evaporation of the solvent necessary to form the solid mesoporous silica spheres leads to an overall size reduction. Therefore, the final particles are approximately half the size of the original droplets. Figure 2A shows a histogram and an SEM image of particles obtained from a polydisperse bulk emulsion. The size distribution is broad and includes a wide range of particles, which is also evident from the image. Using microfluidics significantly reduces the polydispersity. The size distribution and the corresponding SEM image of the particles are shown in Figure 2B. The MFFD used can produce droplets between 25 and 75 μm and particles of about half that size. The particle size distribution in Figure 2B exhibits one well-defined peak centered at around 23 μm. Particles like these are obtained from droplets that were

37

initially between 35 and 40 μm in diameter. After the solvent (water and ethanol) is expelled, the final size is the one shown in Figure 2B. There is a second peak at 30 μm that is most likely due to some coalescence that occurred before the droplets converted to silica particles.

Figure 3 shows a transmission electron microscope image of silica particles with internally ordered mesoporous structure. The particles were selected from the lower end of the distribution curve in Figure 2B. This allowed us to get a better image of the internal mesoporous structure. The pores have an approximately uniform size, which for the surfactant that we used is 6.4 nm, and a pore volume of 0.56 cm3/g. The pores seem to be closed at the surface, which is also the case reported by other authors.

We have engineered microparticles with biporous internal structure.[23] The two types of pores are due to the coexistence of oil microemulsion droplets with smaller ionic micelles in aqueous drops of a few micrometers in diameter. This has been accomplished by tuning the phase state of the oil/water/surfactant mixture to ensure the presence of the different species. In our experiments, large aqueous drops contain smaller microemulsion oil droplets and surfactant micelles. The microemulsion droplets and the micelles in the space between them were used to

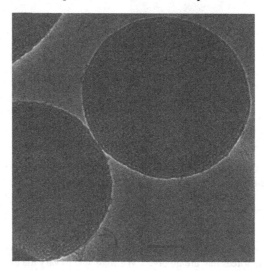

Figure 3. Transmission electron microscope image of silica microspheres containing mesostructured pores that are well ordered by the P104 surfactant. The scale bar is 100 nm. Reproduced from Ref. 22.

template and synthesize silica particles with bimodal porosity. Our procedure used an aqueous tetraethylorthosilicate (TEOS) precursor solution/hexadecane mixture and a combination of nonionic and cationic surfactants [Abil EM90 and cetyltrimethylammonium bromide (CTAB)]. The nonionic surfactant (Abil EM90) is soluble only in the oil phase, while the cationic CTAB is dissolved only in the aqueous phase. Stirring the entire system leads to the formation of micrometer sized aqueous emulsion drops (containing TEOS) dispersed in a hexadecane oil phase. These large drops are stabilized by the oil soluble Abil EM90 surfactant. At the same time, microemulsion oil droplets spontaneously form at the larger aqueous drop interface and occupy its internal volume. This is due to the adsorption of the two surfactants at the oil-water interface and a synergistic drop in the interfacial tension, which facilitates the microemulsion formation.[20, 21, 24-26] Because the CTAB is above the critical micellization concentration (CMC), the aqueous phase will also contain micelles. Finally, the drops are subject to solvent removal and silica polymerization, which fossilizes the microemulsion and micellar structures, producing a bimodal porous network within the microparticles. As the solvent is removed, the micelles may undergo further structural changes and form hexagonal structures.[14, 15] The bimodal pore size distribution is a result of the coexistence of micelles and microemulsion droplets.

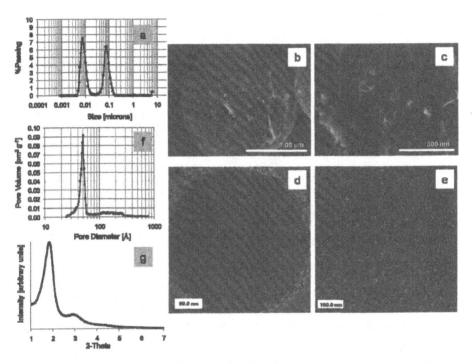

Figure 4. Characterization of the oil/aqueous phase/surfactants solution and the obtained silica particles. (a) DLS measurements of the aqueous phase indicate the presence of micelles (~6 nm) and microemulsion droplets (tens of nm). (b, c) SEM images of silica particle surface with cavities (diameter ~40 nm) arranged as honeycomb. (d) TEM images of silica particle showing open access to internal porous network. (e) Cross section of the silica particle showing microemulsion-templated pores (10-30 nm in diameter) along with smaller, micelle-templated pores (~5 nm in diameter). (f) NLDFT analysis of nitrogen adsorption isotherm suggesting a bimodal pore distribution. (g) XRD pattern verifying the existence of hexagonally packed, micelle-templated pores. Reproduced from Ref. 23.

The coexistence of micelle and microemulsion structures was verified by dynamic light scattering (DLS) analysis of the precursor TEOS/CTAB aqueous phase which was in contact for 48 h with the oil phase containing Abil EM 90. During that time, we observed the spontaneous formation of microemulsion droplets at the interface. The DLS results are shown in Figure 4a where two well-defined peaks are present. The left peak represents the CTAB micelles, while the right peak is due to the microemulsion droplets. DLS measurements of the oil phase showed a single peak with maximum at ~30nm and a long tail toward larger sizes.

The obtained silica particles exhibit a honeycomb-like structure observed by scanning electron microscopy (SEM) (Figure 4b and c), which shows the presence of cavities at the surface with diameters of about 40 nm. A transmission electron microscopy (TEM) micrograph of the particles indicates open access to the porous network (Figure 4d), which facilitates impregnation of the interior with replica materials. Open access to the pores is often not the case if only small surfactant micelles are templated. The TEM image of a particle cross section in Figure 4e and the SEM image of a fractured particle in Figure 5 confirm the presence of the large pores in the interior. The pore size analysis (Figure 4f) suggests the presence of larger

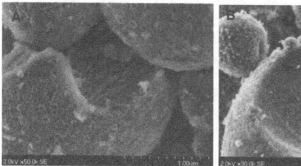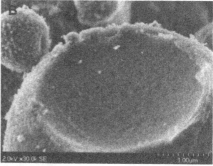

Figure 5. SEM images show the presence of large pores throughout the particles.

(~10-30 nm) and smaller (~5 nm) pores. The cavities at the particle surfaces (observed by SEM) are larger than the pores in the interior of the particle. The obtained silica microspheres have a Brunauer-Emmett-Teller (BET) surface area and pore volume of ~1000 m^2/g and 1.098 cc/g, respectively. Most of the surface area is attributed to the presence of the smaller pores. The smaller pores are not visible in the SEM images (Figure 4c and Figure 5); however, they are detectable by TEM (Figure 4e), adsorption measurements (Figure 4f), and powder X-ray diffraction (XRD) (Figure 4g). They are due to templating of CTAB micelles that are present in the aqueous phase (see the first peak in Figure 4a).

It has been determined that slight variations in microemulsion mixture components (electrolyte concentration, wt% of surfactants, oil to sol ratio, etc.) can produce strikingly different pore morphologies and particle surface areas (see Figure 6). Continuing research focuses on microemulsion kinetic and surfactant adsorption studies to better understand micellar and microemulsion phase formation to enable tailoring of hierarchical pore morphologies and size distributions to specific engineering applications.

40

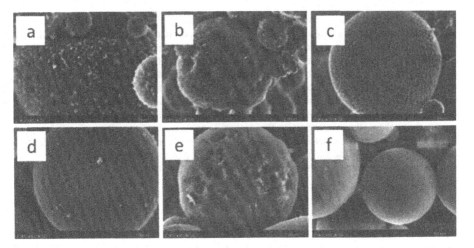

Figure 6. Different pore morphologies and particle surface areas created by varying microemulsion mixture components. Particles formed by: (a) Standard microemulsion mixture, BET surface area 1000 m^2/g. (b) Adding 0.075 M NaCl, BET surface area 650 m^2/g. (c) Adding 0.15 M NaCl, BET surface area 850 m^2/g. (d) 2:1 oil to water ratio, BET surface area 1038 m^2/g. (e) Reducing mass of Abil EM90 surfactant by 25%, BET surface area 975 m^2/g. (f) Increasing mass of CTAB surfactant by 25%, BET surface area 1250 m^2/g.

CONCLUSIONS

We have demonstrated that microfluidics can be successfully used to fabricate monodisperse mesoporous silica particles with well-defined size. The processing conditions for the surfactant templating described in this work are very different from those used in the alternative aerosol method.[27] The kinetics of solvent evaporation in our case is much slower because the solvents (water and alcohol) are transported across the continuous hexadecane phase. This means that the surfactant has a longer time to self-assemble into micellar structures that are closer to equilibrium. In the aerosol method, the solvent removal is very quickly, and some of the structures can be kinetically trapped. Thus, the slower kinetics will lead to better control of the pore structure and size. It also allows the study of the surfactant self-assembly process in silica, which is not tractable when solvent evaporation is too fast. Solvent evaporation and removal across the continuous organic (hexadecane) phase can also be used to obtain other mesoporous structures such as films and membranes. These structures do not require the use of microfluidics but may benefit from the slower kinetics.

41

There are two necessary conditions for obtaining bimodal porous structures through the microemulsion templating procedure described in this paper. First, the two surfactants should significantly decrease the interfacial tension when adsorbed, leading to a spontaneous formation of microemulsion droplets. These droplets are templated to give the larger pores with dimensions of tens of nanometers. Second, the ionic surfactant should be soluble only in the aqueous phase while the nonionic surfactant should be soluble only in the oil phase. The excess ionic surfactant that remains in the bulk forms micelles that are trapped in the solidifying silica, creating the subset of smaller pores with dimensions of a few nanometers. Better understanding of the physics that governs the microemulsion structure will enable the next generation of hierarchically structured particles through controlled design of the phase and templating. The current research aims at tailoring the morphology, pore properties, and functionality of microemulsion-templated materials for application as reduction and oxidation electrocatalysts in fuel cells.

ACKNOWLEDGMENTS

This work was supported by NSF (PREM/DMR 0611616), DoE-EERE, V.C.7 Advanced Cathode Catalysts, DoE-EPSCoR Implementation Program: Materials for Energy Conversion, and the W. M. Keck Foundation.

REFERENCES

1.   Nie, Z.H., et al., *Polymer particles with various shapes and morphologies produced in continuous microfluidic reactors.* Journal of the American Chemical Society, 2005. **127**(22): p. 8058-8063.
2.   Dendukuri, D., et al., *Controlled synthesis of nonspherical microparticles using microfluidics.* Langmuir, 2005. **21**(6): p. 2113-2116.
3.   Anna, S.L., N. Bontoux, and H.A. Stone, *Formation of dispersions using "flow focusing" in microchannels.* Applied Physics Letters, 2003. **82**(3): p. 364-366.
4.   Xu, S., et al., *Generation of monodisperse particles by using microfluidics: Control over size, shape, and composition (vol 44, pg 724, 2005).* Angewandte Chemie-International Edition, 2005. **44**(25): p. 3799-3799.
5.   Ikkai, F., et al., *New method of producing mono-sized polymer gel particles using microchannel emulsification and UV irradiation.* Colloid and Polymer Science, 2005. **283**(10): p. 1149-1153.
6.   Serra, C., et al., *A predictive approach of the influence of the operating parameters on the size of polymer particles synthesized in a simplified microfluidic system.* Langmuir, 2007. **23**(14): p. 7745-7750.
7.   Zhang, H., et al., *Microfluidic production of biopolymer microcapsules with controlled morphology.* Journal of the American Chemical Society, 2006. **128**(37): p. 12205-12210.
8.   Cejka, J. and S. Mintova, *Perspectives of micro/mesoporous composites in catalysis.* Catalysis Reviews-Science and Engineering, 2007. **49**(4): p. 457-509.

9.   Hartmann, M., *Ordered mesoporous materials for bioadsorption and biocatalysis.* Chemistry of Materials, 2005. **17**(18): p. 4577-4593.

10.   Gasteiger, H.A., et al., *Activity benchmarks and requirements for Pt, Pt-alloy, and non-Pt oxygen reduction catalysts for PEMFCs.* Applied Catalysis B-Environmental, 2005. **56**(1-2): p. 9-35.

11.   Gallis, K.W., et al., *The use of mesoporous silica in liquid chromatography.* Advanced Materials, 1999. **11**(17): p. 1452-1455.

12.   Vallet-Regi, M., et al., *A new property of MCM-41: Drug delivery system.* Chemistry of Materials, 2001. **13**(2): p. 308-311.

13.   Kresge, C.T., et al., *Ordered Mesoporous Molecular-Sieves Synthesized by a Liquid-Crystal Template Mechanism.* Nature, 1992. **359**(6397): p. 710-712.

14.   Lu, Y.F., et al., *Aerosol-assisted self-assembly of mesostructured spherical nanoparticles.* Nature, 1999. **398**(6724): p. 223-226.

15.   Brinker, C.J., et al., *Evaporation-induced self-assembly: Nanostructures made easy.* Advanced Materials, 1999. **11**(7): p. 579-585.

16.   Bore, M.T., et al., *Hexagonal mesostructure in powders produced by evaporation-induced self-assembly of aerosols from aqueous tetraethoxysilane solutions.* Langmuir, 2003. **19**(2): p. 256-264.

17.   Andersson, N., et al., *Combined emulsion and solvent evaporation (ESE) synthesis route to well-ordered mesoporous materials.* Langmuir, 2007. **23**(3): p. 1459-1464.

18.   Schacht, S., et al., *Oil-water interface templating of mesoporous macroscale structures.* Science, 1996. **273**(5276): p. 768-771.

19.   Huo, Q.S., et al., *Preparation of hard mesoporous silica spheres.* Chemistry of Materials, 1997. **9**(1): p. 14-17.

20.   Hu, Y. and J.M. Prausnitz, *Molecular Thermodynamics of Partially-Ordered Fluids - Microemulsions.* Aiche Journal, 1988. **34**(5): p. 814-824.

21.   Nagarajan, R. and E. Ruckenstein, *Molecular theory of microemulsions.* Langmuir, 2000. **16**(16): p. 6400-6415.

22.   Carroll, N.J., et al., *Droplet-based microfluidics for emulsion and solvent evaporation synthesis of monodisperse mesoporous silica microspheres.* Langmuir, 2008. **24**(3): p. 658-661.

23.   Carroll, N.J., Pylypenko, P., Atanassov, P.B., and Petsev, D.N. , *Microparticles with Bimodal Nanoporosity Derived by Microemulsion Templating.* Langmuir, 2009.

24.   Aveyard, R., et al., *Interfacial-Tension Minima in Oil-Water Surfactant Systems - Behavior of Alkane Aqueous Nacl Systems Containing Aerosol Ot.* Journal of the Chemical Society-Faraday Transactions I, 1986. **82**: p. 125-142.

25.   Aveyard, R., B.P. Binks, and J. Mead, *Interfacial-Tension Minima in Oil + Water + Surfactant Systems - Effects of Salt, Temperature and Alkane in Systems Containing Ionic Surfactants.* Journal of the Chemical Society-Faraday Transactions I, 1985. **81**: p. 2169-2177.

26.   Schmidt-Winkel, P., C.J. Glinka, and G.D. Stucky, *Microemulsion templates for mesoporous silica.* Langmuir, 2000. **16**(2): p. 356-361.

27.   Rao, G.V.R., et al., *Monodisperse mesoporous silica microspheres formed by evaporation-induced self assembly of surfactant templates in aerosols.* Advanced Materials, 2002. **14**(18): p. 1301-1304.

43

Mater. Res. Soc. Symp. Proc. Vol. 1272 © 2010 Materials Research Society   1272-KK09-04

Numerical Simulations of Misalignment Effects In Microfluidic Interconnects

Sudheer D Rani[1,3], Taehyun Park[1,3], Byong Hee You[1,4], Steven A. Soper[2,3],

Michael C. Murphy[1,3] and Dimitris E. Nikitopoulos[1,3]

[1]Department of Mechanical Engineering, Louisiana State University, Baton Rouge, LA 70803, U.S.
[2]Department of Chemistry, Louisiana State University, Baton Rouge, LA 70803, U.S.
[3]Center for Bio-Modular Multi-Scale Systems, Louisiana State University, Baton Rouge, LA 70803, U.S.
[4]Department of Engineering Technology, Texas State University-San Marcos, 601 University Drive, San Marcos, TX 78666, U.S.

ABSTRACT

Numerical simulations were performed to see the effect of geometrical misalignment in pressure driven flows. Geometric misalignment effects on flow characteristics arising in three types of interconnection methods a) end-to-end interconnection, b) channel overlap when chips are stacked on top of each other, and c) the misalignment occurring due to the offset between the external tubing and the reservoir were investigated. For the case of end-to-end interconnection, the effect of misalignment was investigated for 0, 13, 50, 58, and 75% reduction in the available flow area at the location of geometrical misalignment. In the interconnection through channel overlap, various possible misalignment configurations were simulated by maintaining the same amount of misalignment (75% flow area reduction) for all the configurations. The effect of misalignment in a Tube-in-Reservoir interconnection was investigated by positioning the tube at an offset of 164μm from the reservoir center. All the results were evaluated in terms of the equivalent length of a straight pipe. The effect of reynolds number (Re) was also taken into account by performing additional simulations of aforementioned cases at reynolds numbers ranging from 0.075 to 75. The results are interpreted in terms of equivalent length (L_e) as a function of Re and misalignment area ratio (A_1:A_2), where A_1 is the original cross-sectional area of the channel and A_2 is the available flow area at mismatch location. Equivalent length calculations revealed that the effect of misalignment in tube-in-reservoir interconnection method was the most insignificant when compared to the other two methods of interconnection.

INTRODUCTION

Recent advances in microfluidics have resulted in the development of miniaturized devices to accomplish vast range of functions ranging from commercial inkjet printing technologies to genetic analysis. Portability, reduced reagent usage, faster analysis times, high throughput analysis are some the major advantages that can rival the current benchtop systems. In addition, the ability to modularize these devices by assembling series of modules or components to perform certain functions is one of the key driving forces in the development of microfluidic devices. Each module can be tested separately with the ability to add additional modules to improve the overall functionality of the microfluidic system. For example, an integrated microfluidic device was recently demonstrated with polymerase chain reaction (PCR) on polycarbonate chip (PC) and Ligase detection reaction (LDR) performed on poly-methyl-methacrylate (PMMA) chip to detect low-abundant DNA mutations in gene fragments (K-ras) with colorectal cancers [1]. One of the most challenging issues in the successful realization of an integrated microfluidic device is the need to provide fluidic interconnection to provide fluidic passages for microfluidic components or modules. Fluidic interconnections have been established using interlocking finger joints for vertically stacked modules [2], or stacking modules by silica capillaries [3], plasma assisted bonding [4] or O-rings [5]. The fluidic interconnects should be designed to withstand high operating pressures, have minimal dead volumes and should be leak free. In order to address the manufacturing and functionality challenges involved in the design of interconnects, it is required to have insights into the possible effects of geometric misalignments in interconnected microchannels. Misalignments have been shown to have adverse effect both terms of pressure drop [6] as well as sample dispersion [7]. In the present work, numerical simulations have been performed to investigate the possible effects of misalignments arising in three different interconnection methods; end to end, interconnection through channel overlap and tube-in-reservoir interconnection.

NUMERICAL SIMULATIONS

Misalignment effects in End-to-End Interconnection

Numerical simulations have been performed using CFD software FLUENT^{TM}. The equations solved are the continuity equation coupled with the momentum equations to resolve the full flow field. The grid was varied between 0.5 million cells to 2 million cells depending on the misalignment configuration and the flow behavior at different Reynolds numbers. The mesh was sufficiently refined and grid convergence checked in all the cases to ensure that the results are grid independent. The simulation domain shown in Figure 1 consists of two channels with rectangular cross section and connected end to end. A geometrical misalignment was introduced

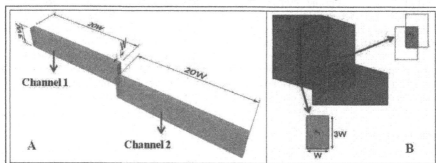

Fig. 1 A) 3D model showing the channels connected end-to-end with the dimensions shown in terms of the width w of the channel. B) Cross sectional area A_1 of the channel before misalignment and the reduced flow area A_2 at misalignment location.

as shown in Figure 1. The cross sectional area o the channel is denoted by A_1 and the available flow area at the mismatch plane is denoted by A_2. Similar cross-sectional area convention (i.e. $A_1 \& A_2$) is employed for all interconnection schemes. The effect of misalignment was investigated for varying flow area ratios ($A_1 : A_2$) of **1 : 1**, **1.14 : 1, 2 : 1, 2.4 : 1** and **4 : 1** where the ratio $A_1 : A_2 = 1:1$ corresponds to the case of no geometrical misalignment and $A_1 : A_2 = 4:1$ corresponds to extreme case of geometrical misalignment (i.e. 75% reduction in the available flow area at misalignment plane). To account for the effect of geometrical misalignment on the flow behavior, each case was run for varying reynolds numbers of 0.075, 15, 25, 37.5 and 75.

Misalignment effects in interconnection through channel overlap

The effect of geometrical misalignment in the case of interconnection through channel overlap has been investigated. This type of misalignment frequently occurs during the alignment of two fluidic chips. For example, the misalignment can be upto 103±6 μm in the direction of X-axis and 16±4 μm in the direction of Y-axis for the case of chips being aligned through V-grooves [8]. The top view of various possible misalignment configurations when fluidic chips are connected through channel overlap interconnection are shown in Figure 2.

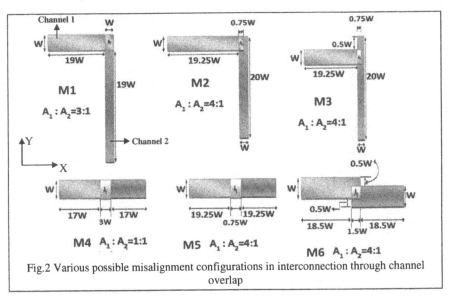

Fig.2 Various possible misalignment configurations in interconnection through channel overlap

The channel overlap configurations are named from M1 to M6 and the channels considered in the present study have a width (w) of 50μm, depth (H) of 150μm and a length (L) of 1000μm. In Figure 2, the dimensions of the channel and the misalignment offset are given in terms of the width (i.e. 1000μm length is depicted as $20\,w$ with width being $w = 50$ μm). In the case of M1, M2 and M3 the axis of the channels are perpendicular while M4, M5 and M6 have their axes parallel to each other. Geometric misalignments were introduced in various amounts along both

47

X and Y axes and were quantified in terms of misalignment area ratio $A_1 : A_2$. For example in M2, Channel 2 is given an offset of $0.25w$ (i.e. $0.25 \times 50 = 12.5 \mu m$ since width $w = 50 \mu m$) in the X-direction, resulting in 75% reduction in the original cross-sectional area. Hence the misalignment area ratio in this case becomes $A_1 : A_2 = 4:1$. Similar geometric misalignments were introduced in all other cases and are depicted in Figure 2. The case M1 has a misalignment area ratio $A_1 : A_2 = 3:1$ while the cases M2, M3, M5 and M6 have an area ratios of $4:1$. In order to quantify the effects of misalignment, these cases are compared with the case of M4, which has been given the maximum available flow area at the mismatch plane i.e. $A_1 : A_2 = 1:1$.

Misalignment effects in Tubing-in-Reservoir interconnection

The misalignment arising between tubing and a reservoir had been dealt in the present study owing to its importance in fluidic interconnection. Often the drilled hole is not located exactly at the center of the reservoir, resulting in an offset from the center. This type of misalignment can also arise during the bonding of the external tubing to the chip. The case of perfect alignment is an ideal one, which is seldomly obtained during interconnection. In order to estimate the amount of misalignment occurring during tube-in-reservoir interconnection, images were taken for two different chips under the microscope with drilled holes and bonded tubing. The offset obtained was as high as 164 µm. In order to investigate such misalignment effects by simulation, two cases with and without the offset of tubing from the center of reservoir were compared. The dimensions of the tubing and reservoir used in the simulation (i.e. 85 µm and 500 µm radii respectively) were comparable to the ones used in the experiment. For all the cases simulated, the fluid in the channel was assumed to have the properties of water.

RESULTS AND DISCUSSION

When an incompressible fluid flows through the misalignment, the abrupt area change causes a drop in pressure at the misalignment plane. This pressure drop at the location of the misalignment can be written in terms of equivalent length of a straight pipe L_e analogous to the drop due to sudden expansion or contraction as [9]

$$h_m = \frac{\Delta P_M}{\rho} = f\left(\frac{L_e}{D_h}\right)\frac{1}{2}\overline{V}^2 \tag{1}$$

where ΔP_M is the pressure drop at the mismatch plane, h_m is the head loss due to the geometrical mismatch and f is the friction factor evaluated in the straight part of the channel. L_e, D_h, $\overline{V}$ are the equivalent length of a straight pipe, the hydraulic diameter and the average velocity in the channel. The nondimensional equivalent length L_e / D_H scaled by $\left[\left(\frac{A_1}{A_2}\right)^2 - 1\right]$ as a function of reynolds number for area ratios $A_1 : A_2 = $ **1.14 : 1, 2 : 1, 2.4 : 1** and **4 : 1** is evaluated for both

48

end to end and channel overlap interconnection methods. This data for scaled equivalent length can be curve fitted to develop a correlation of equivalent length $L_e = L_e\left(Re, \dfrac{A_1}{A_2}\right)$.

For example Figure 4 shows the scaled equivalent length as a function of reynolds number for the case of interconnection through channel overlap. It can be seen that the equivalent length can be more than 20 hydraulic diameters for the extreme case of $A_1 : A_2 = 4:1$ and Re=75, due to the increased viscous dissipation at high Reynolds numbers. The high drop in pressure at the misalignment location is a combination of pressure drop due to sudden contraction and pressure drop due to viscous losses. The recirculation region shown for a particular case of interconnection through channel overlap in Figure 5 corroborates the additional pressure drop due to viscous dissipation. The interesting thing to note from Figure 4 is that, the pressure drop at misalignment is independent of the configuration as almost all the curves overlap on top of each other. Based on this unique phenomena, correlations can be developed to predict the pressure drop for a given misalignment ratio and reynolds number. Similar calculations of equivalent length are carried out for the case of tube-in-reservoir interconnection and are not shown here.

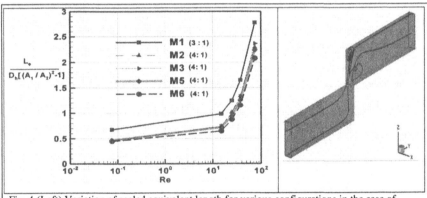

Fig. 4 (Left) Variation of scaled equivalent length for various configurations in the case of interconnection through channel overlap.
Fig. 5 (Right) Streamlines depicting recirculation regions for configuration M6 at Re = 75 due to the misalignment

CONCLUSIONS

The effect of geometrical misalignment through numerical simulations has been investigated for three types of interconnection methods: End-to-End interconnection, interconnection through channel overlap and interconnection through Tube-in-reservoir. There was a drastic increase in the pressure drop as the available flow area decreased and this was found to be enhanced at higher Reynolds numbers due to viscous dissipation. Based on the calculations of equivalent length it can also be stated that the effect of misalignment becomes pronounced for flow area ratios greater than $A_1: A_2 = 2 : 1$. It was found that the effect of tube-in-reservoir was insignificant when compared to interconnection through end-to-end and channel overlap.

ACKNOWLEDGEMENT

This work was supported by a Bioengineering Research Partnership (NIH R24-EB002115) through the National Human Genome Research Institute (NHGRI), the National Cancer Institute (NCI) of the National Institutes of Health (NIH), and partially funded by National Science Foundation under Grant EPS-0346411 and the State of Louisiana Board of Regents Support Fund.

REFERENCES

[1]   M. Hashimoto, F. Barany, and S. A. Soper, "Polymerase chain reaction/ligase detection reaction/hybridization assays using flow-through microfluidic devices for the detection of low-abundant DNA point mutations," *Biosensors and Bioelectronics,* vol. 21, pp. 1915-1923, 2006.

[2]   C. Gonzalez, S. D. Collins, and R. L. Smith, "Fluidic interconnects for modular assembly of chemical microsystems," *Sensors and Actuators B: Chemical,* vol. 49, pp. 40-45, 1998.

[3]   G. Blankenstein and U. Darling Larsen, "Modular concept of a laboratory on a chip for chemical and biochemical analysis," *Biosensors and Bioelectronics,* vol. 13, pp. 427-438, 1998.

[4]   O. Hofmann, P. Niedermann, and A. Manz, "Modular approach to fabrication of three-dimensional microchannel systems in PDMS—application to sheath flow microchips," *Lab Chip,* vol. 1, pp. 108-114, 2001.

[5]   G. Perozziello, F. Bundgaard, and O. Geschke, "Fluidic interconnections for microfluidic systems: A new integrated fluidic interconnection allowing plug`n'play functionality," *Sensors and Actuators B: Chemical,* vol. 130, pp. 947-953, 2008.

[6]   P. Aniruddha and H. A. Chong, "Self-aligning microfluidic interconnects for glass- and plastic-based microfluidic systems," *Journal of Micromechanics and Microengineering,* vol. 12, p. 35, 2002.

[7]   S. D. Rani, S. A. Soper, D. E. Nikitopoulos, and C. M. Murphy, "Simulation of Electroosmotic and Pressure Driven flows in Microfluidic Interconnects," in *ASME Conference Proceedings (IMECE 2006),Paper no: IMECE2006-15388,459-465,* Chicago, Illinois, 2006, pp. 459-465.

[8]   H. B. You, C. P. chen, W. J. Guy, P. Datta, D. E. Nikitopoulos, S. A. Soper, and M. C. Murphy, "Passive Alignment Structures in Modular, Polymer Microfluidic Devices," in *IMECE,* 2006.

[9]   F. F. Abdelall, G. Hahn, S. M. Ghiaasiaan, S. I. Abdel-Khalik, S. S. Jeter, M. Yoda, and D. L. Sadowski, "Pressure drop caused by abrupt flow area changes in small channels," *Experimental Thermal and Fluid Science,* vol. 29, pp. 425-434, 2005.

Mater. Res. Soc. Symp. Proc. Vol. 1272 © 2010 Materials Research Society 1272-KK09-05

Hot Embossing of Microfluidic Channel Structures in Cyclic Olefin Copolymers

Patrick W. Leech[1,] Xiaoqing Zhang[1] and Yonggang Zhu[2]
[1]CSIRO Materials Science and Engineering, Clayton, 3168, Victoria, Australia
[2]CSIRO Materials Science and Engineering, Highett, 3193, Victoria, Australia

ABSTRACT

The dynamic mechanical behaviour of a series of cyclic olefin copolymers (COCs) with varying norbornene content has been examined in the vicinity of the glass transition temperature, T_g. The temperature of the transition has been shown to increase linearly with increase in norbornene content. Measurements of both the elastic storage modulus, E', and loss modulus, E'', have decreased exponentially with rise in temperature above T_g. A levelling-off in E'' occurred at ≥ 20 °C above T_g for all copolymers. The results of Dynamic Mechanical Thermal Analysis (DMTA) have been used in the identification of optimum conditions for hot embossing. At ≥ 20 °C above T_g in a region of viscous liquid flow, the hot embossing of COC has resulted in a full replication of channel depth without cracking or distortion.

INTRODUCTION

Cyclic olefin copolymer (COC) has emerged in recent years as an attractive substrate for use in microfluidic devices. COC has exhibited a unique combination of properties including an optical transparency extending into the DUV range, chemical inertness and negligible moisture absorption. Another attribute of COC has been the ability to tailor it's thermal and mechanical properties by variation in the ratio of cyclic monomer (typically norbornene) to olefin (ethylene) [1]. In copolymers with $\leq$ 40% norbornene, the chemical structure of COC has been identified by nuclear magnetic resonance as comprising sequences of norbornene and ethylene units within the backbone of the main polymer chain [2]. Higher norbornene contents have acted to stiffen the main chain through the substitution of ethylene units by the bulky ring structure. As a consequence, an increase in norbornene content in COC has correlated with a linear rise in glass transition temperature, T_g [3], an increase in microhardness [4] and tensile strength [4] and decrease in ductility [5]. The linear dependence of T_g on composition was shown by Forsyth et.al. [6] as sensitive to the type of microstructure within the copolymer at high norbornene content.

COC has recently been used in the replication of microfluidic devices by processes of hot embossing [7,8] and injection molding [9,10]. In these studies, a number of different compositions of copolymer have been applied in the fabrication of channel structures. However, little systematic data exists on the effect of comonomer content in the replication of micro-patterns in COC. This paper has examined both the dynamic mechanical behaviour and the hot embossing properties of a series of standard grades of COC. The norbornene/ ethylene compositions of the copolymers were selected as providing a wide range of T_g. All grades have been synthesised using a common metallocene catalyst. Measurements of thermal deformation properties of the copolymers have been combined with the analysis of hot embossing characteristics. The temperature range in the vicinity of the glass transition has been selected for these experiments as widely used in the replication of microstructures during hot embossing [12]. The results of experiments on the optimisation of embossing parameters have been applied in the fabrication of microfluidic channels.

EXPERIMENTAL DETAILS

The six grades of Topas COC used in these experiments have been listed in Table 1. The copolymers contained a range of norbornene contents (61-82 wt%) as supplied by Polyplastics, Japan. The level of incorporation of norbornene in COC has been reported by Forsyth et al. as directly dependent on the ratio of norbornene/ ethylene [6]. The six grades have been designated as C-61 to C-82 corresponding to the wt % norbornene. Dynamic Mechanical Thermal Analysis (DMTA) was performed on samples using a PerkinElmer PYRIS™ system in dual cantilever bending mode. During DMTA, a plate specimen with dimensions of 10 x 100 x 1 mm was heated at a rate of 2.0 °C min^{-1} while applying a sinusoidally varying strain. The elastic storage modulus, E', loss modulus, E", and loss factor, tan δ, were simultaneously determined at frequencies of 0.1-20 Hz. Experiments on the hot embossing of the series of copolymers were performed using a hydraulic press with platens of brass and a nickel shim as a die. The pattern of the channels was fabricated in dry film resist (Shipley 5038) in a collimated UV system at 16.5 W/cm^2 (λ = 350-450 nm). The master pattern in resist was replicated as a Ni shim (hardness of ~265 HV) using an initial sputter deposition of 100 nm followed by electroplating in a nickel suphamate bath to a thickness of 150 nm. Hot embossing of the channel pattern was performed at a temperature of 80-180 °C and a force of 25 kN.

Table I. Samples of COC (Topas) with wt % Norbornene

Sample	C-61	C-65	C-75	C-76	C-79	C-82
Norbornene wt%	61	65	75	76	79	82
COC Grade	9506F-04	8007S-04	5013L-10	6013X12	6015S-04	6017S-04

RESULTS AND DISCUSSION

Dynamic Mechanical Thermal Analysis

Fig. 1 shows plots of E' and tan δ versus temperature in the range 20-200 °C. Each curve in Fig. 1(a) has exhibited a rapid decrease in E' above a critical temperature. A single α relaxation of this type has previously been attributed to the glass transition in COC [4,6]. T_g was also evident as maxima in tan δ in Fig. 1(b). The temperature corresponding to the maximum in tan δ in Fig. 1(b) increased directly with % norbornene, although the intensity of the maximum remained constant at tan δ ~1.4 in all grades. Fig. 2 shows the temperature of the α relaxation plotted as a function of % norbornene. In Fig. 2, the primary α relaxation was measured by the peak temperature in tan δ and the end-point in E'. This trend of increase in T_g with norbornene content (Fig. 2) has been attributed to the increased stiffness in the copolymer chain with a greater density of cyclic monomer units [6].

At room temperature, the value of the elastic storage modulus, E', was measured in the range 2,300-3,000 x 10^6 Pa. At 40 °C above the transition, E' had decreased to a value of 0.2 - 1.6 x 10^6 Pa, a drop in modulus of more than 3 orders of magnitude. The continuous and rapid decrease in E' which was evident at temperatures above T_g and the slight change in

52

slope in Fig. 1(a) has indicated that all of the grades of COC had transformed directly from the glass transition region to a regime of viscous liquid flow. This regime has previously been characterised by an irreversible deformation of the polymer together with a storage modulus of $E' \sim 10^5$-10^6 Pa, with the value of E' depending on polymer species and concentration [11,12]. The deformation in this regime has been associated with a mechanically activated movement of polymer chain segments.

Figs. 3(a) and (b) show the variation in E' and E'' in the range of temperature immediately above T_g. Fig. 3(a) shows that the value of E' decreased exponentially above T_g for each of the respective grades. At any given temperature above T_g, the range in E' was within a band which varied by approximately an order of magnitude between the grades.

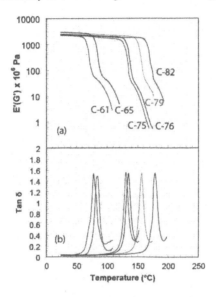

Figure 1. E' and Tan δ versus temperature obtained from DMTA (1 Hz).

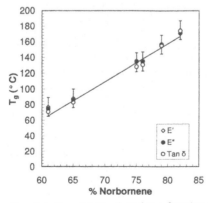

Figure 2. Temperature of α relaxation at 1 Hz plotted as a function of % Norbornene.

53

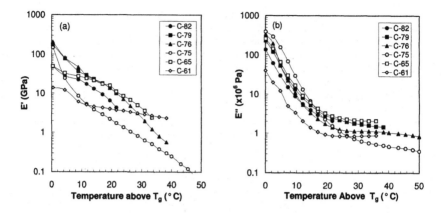

Figure 3. (a) E' and (b) E" at 1 Hz versus °C above the transition for each grade of COC.

In Fig. 3(b), the loss modulus, E" decreased exponentially with increase in temperature until levelling off at $\geq$ 20 °C above T_g. The near constant value of E" with increasing temperature at $\geq$20 °C above T_g has indicated that the viscous deformation was essentially independent of temperature in this range. The results of DMTA in Figs. 1-3 have shown an optimal range of temperature at $\geq$ 20 °C above the transition for embossing of all grades of COC. In this range, a negligibly small value of elastic modulus, E', was combined with a low and temperature invariant value of loss modulus, E". This combination of parameters has enabled the hot embossing of COC without residual elastic deformation while maintaining a low energy loss due to viscous liquid flow. A temperature range of 20-40 °C above the transition has been used empirically in the hot embossing of polymers [12].

Hot Embossing Experiments

Fig. 4 shows SEM images and profilometry traces of 200 μm wide channels which were embossed in C-76 grade. The channels were embossed at the lower, mid-point and upper temperature in the transition range. At the lower temperature limit of the transition (80 °C), Fig. 4(a) shows a broad region of edge roughness along the channel together with evidence of delamination. The corresponding profile has shown a convex distortion at the base of the channel due to uneven elastic recovery following viscous flow. Fig. 4(b) shows a reduction in the width of the edge roughness combined with an increase in the embossed depth. At the upper limit of the transition temperature at 140 °C, Fig. 4(c) shows a smooth sidewall with a narrow region of deformed edge. The depth of embossing was significantly greater at this temperature than below the transition, with formation of an essentially flat base in the channel. A slight ridge was present (see profile in Fig. 4(c)) due to displacement of polymer on either side of the channel. A similar pattern of deformation as a function of temperature was also evident in the other grades of COC. However, in COC grades with higher norbornene content (C-79 and C-82), the formation of multiple cracks was evident along the length of the channel after embossing at temperatures below the transition. Cracks of this type were absent during embossing at temperatures above T_g. The formation of fibrillated crazes has been previously identified during deformation of COC containing higher levels of

% norbornene (68-78 wt%) [4]. A reduction in ductility of COC has been correlated with increase in % norbornene [4].

Figs. 5 (a) and (b) show plots of the depth of embossing in 200 μm wide channels versus the duration of loading and process temperature, respectively. In Fig. 5(a), the curve for each grade has shown a steep rise in depth with increase in embossing time until reaching an approximately constant value at ~60 s. The maximum depth of 70 μm was equivalent to the height of the ridge in the nickel shim. Fig. 5(a) shows that, at embossing times of ≤ 60 s, a wide variation in depth was evident between the grades of COC. The embossed depth increased with a reduction in % norbornene. Based on Fig. 5(a), an embossing time of 120 s was used as a standard procedure for the experiments.

Fig. 5(b) shows that the embossed depth increased sharply over a critical range of temperature for all COC grades. At temperatures below the critical transition range, the depth of embossing was approximately constant at 40-50 μm. At temperatures above the critical range, the embossed depth increased to a plateau level of ~70 μm (the ridge height in the shim). In general, the range of transition temperature was detected at a lower temperature in hot embossing experiments than DMTA. In comparison, the upper limit of the transition was similar for both embossing experiments and DMTA.

All embossing experiments which were performed in the vicinity of the transition temperature have resulted in a permanent deformation. However, a temperature above T_g was required for the full definition of the channel without a distortion of the COC due to residual elastic deformation. In the range of temperature above T_g, both the elastic modulus, E', and loss modulus, E'', had decreased to 1-10 x 10^6 Pa while E'' was approximately constant with temperature. In this region of viscous liquid flow above T_g, the deformation during hot embossing was less sensitive to variation in temperature than within the transition range.

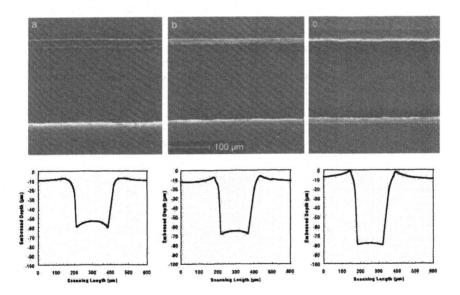

Figure 4. Scanning electron micrographs and profilometry traces of channels embossed in 6013X12 grade at a) 80 °C, (b) 110 °C and (c) 140°C at 25 kN force and 120 s duration.

55

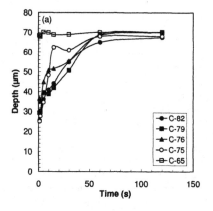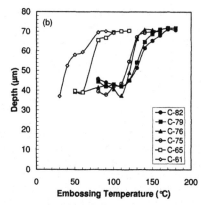

Figure 5. Embossed depth versus (a) time at 170 °C and 25 kN force and b) temperature at 25 kN and 120 s.

CONCLUSIONS

Dynamic mechanical thermal analysis of a series of cyclic olefin copolymers has shown a linear dependence of T_g on norbornene content. At temperatures above the transition in each grade, the magnitude of the elastic storage modulus, E', has decreased exponentially with temperature. At 40 °C above the transition, the elastic modulus E' had decreased to 0.2-1.6 x 10^6 Pa, more than 3 orders of magnitude below E' at room temperature. The loss modulus, E", has also been shown to decrease sharply with temperature before levelling-off at $\geq$ 20 °C above T_g. In this range of temperature, the deformation of COC was characterised by a low values of E' in combination with low and temperature invariant values of E". Hot embossing of microchannels in this regime of viscous liquid flow has produced smooth sidewalls, a flat base and full depth.

REFERENCES

1. D. McNally *in Encyclopedia of Polymer Science and Technology*, Vol.2, 3rd Ed, H.F. Mark Ed., Wiley, Hoboken NJ, 489 (2003).
2. J.Y. Shin, J.Y. Park, C. Liu, J. He and S.C. Kim, *Pure Appl.Chem.* **77(5)**, 801 (2005).
3. J.F. Forsyth, T. Scrivani, R. Benavente, C. Marestin and J.M. Perena, *J.Appl.Polymer Sci.* **82**, 2159 (2001).
4. T. Scrivani, R. Benaventure, E. Perez and J.M. Perena, *Macromol.Chem.Phys.* **202**, 2547 (2001).
5. V. Seydewitz, M. Krumova, G.H. Michler, J.Y. Park and S.C. Kim, *Polymer* **46**, 5608 (2005).
6. J.F. Forsyth, J.M. Perena, R. Benavente, E. Perez, I. Tritto, L. Boggioni and H-H. Brintzinger, *Macromol.Chem.Phys.* **202,** 614 (2001).
7. N.S. Cameron, H. Roberge, T. Veres, S.C. Jakeway and H.J. Crabtree, *Lab Chip* **6**, 936 (2006).
8. J. Steigert, S. Haeberle, T. Brenner, C. Müller, C.P. Steinert, P. Koltay, N. Gottschlich, H. Reinecke, J. Rühe, R. Zengerle and J. Ducré, *J.Microm.Microeng.* **17**, 333 (2007).
9. C-C Hong, J-W. Choi and C.H. Ahn, *Lab.Chip.* **4**, 109 (2004).
10. D.A. Mair, E. Geiger, A.P. Pisano, J.M.J. Fréchet and F. Svec, *Lab Chip* **6**, 1346 (2006).
11. X. Shan, Y.C. Liu and Y.C. Lam, *Microsyst.Technol.* **14**, 1055 (2008).
12. L.J. Guo, *Journal of Physics D:Applied Physics*, 37, R123 (2004).

Mater. Res. Soc. Symp. Proc. Vol. 1272 © 2010 Materials Research Society   1272-KK09-06

Toward a Merged Microfluidic/Microelectrode Array Device for Culture of Unidirectional Neural Network.

Alexander H. Mo[1] and Sarah C. Heilshorn[1]
[1]Department of Materials Science and Engineering, Stanford University, Stanford, CA 94305-4045, U.S.A.

ABSTRACT

Traditional neural cultures are formed from disassociating primary neurons that are sourced from animals. However by disassociating them, any larger organizational structure between the neurons is lost. In an effort to regain some of the lost organization a device that is capable of recreating and monitoring a unidirectional connection between two populations is proposed. Initial validation toward a fully functional device is presented. Using soft lithography techniques, a microfluidic chip has been designed and produced with a design conducive for facilitating such a neuron culture. Using a specialized adhesive method, this chip can be attached and removed from a commercially available microelectrode array. Finally, cell viability is demonstrated using the PC12 neuron-like cell line after exploring several device configurations. Further efforts will be focused on primary neuron culture and establishment of a unidirectional network.

INTRODUCTION

The brain is a centralized collective of neurons organized into nodes that are responsible for processing stimuli and decision making in higher organisms. Entire branches of science are dedicated to understanding the different organizational levels that make up the brain from high-level consciousness to communication between two neurons. However on the cellular level, neuron cultures typically use disassociated neurons and it becomes difficult to study organizational behavior these neurons once had. The development of soft lithography offers the tools to recreate some of the organization that was lost with disassociated neurons.

Toward this end, we report progress toward developing a device that can create a unidirectional connection between two otherwise isolated populations of randomly networked neurons and monitor electrical communication between these populations. This is accomplished by merging a microfluidic chip onto a commercially available microelectrode array. The microfluidic chip is made from the cheap, biocompatible elastomer polydimethylsiloxane (PDMS). Figure 1 shows the device design containing two cell chambers connected by small axon guidance channels modeled on similar devices.[1] These chambers are aligned over the electrodes of a commercially available microelectrode array and bonded with a special method to create the finished device.

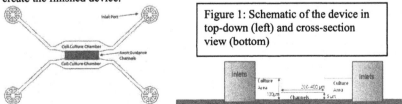

Figure 1: Schematic of the device in top-down (left) and cross-section view (bottom)

Device validation work involving device assembly and cell viability are reported here. Methods for bonding and removing microfluidic chips from microelectrode arrays have been developed. Cell viability in microfluidic devices is tested using a neuron-like cell line called PC12. Due to the iterative nature of the work, microfluidic chips were bonded to inexpensive microscope slides instead of expensive microelectrode arrays. Different device configurations were explored to find the optimal one for cell viability. Daily phase contrast microscopy was used to inspect overall health of cultures in devices. A cell viability assay was performed to qualitatively confirm long term cell viability in the optimal configuration.

EXPERIMENT

Microfluidic Device Fabrication

The microfluidic component of the merged device was fabricated from PDMS using soft lithography techniques reported by Xia, et al. [2] Device designs were drawn using AutoCAD (Autodesk, San Rafael, CA) and sent to the Stanford University Microfluidics Foundry (SUMF) for production. SUMF patterned negative reliefs of the devices on to silicon wafers with SU-8 photoresist in order to create the master molds.

After completion of the molds, the master mold was placed in a wafer storage box along with plastic dish containing 10 mL of chlorotrimethylsilane (Sigma-Aldrich, St. Louis, MO). The chlorotrimethylsilane was allowed to evaporate from the dish onto the master mold over the course of 1 hr in order to facilitate the removal of PDMS from the master mold. Afterward the master mold was placed in dish folded from aluminum foil. PDMS (Sylgard 184, Dow Corning, Midland , MI) was mixed together in a 10:1 weight ratio of prepolymer to curing agent and cast into the dish. The aluminum foil dish was then placed in a vacuum desiccator for 30 minutes to degas the PDMS. Afterward the aluminum foil dish with the master mold and degassed PDMS was placed in a dry oven for 1 hr at 70°C.

After separating the hardened PDMS from the master mold, individual chips were cut out from the PDMS block and punches (Technical Innovations, Angleton, TX) were used to make holes for fluid inlets. Different device configurations employ 2.7-mm and 0.5-mm diameter holes punched in different locations (Figure 2)

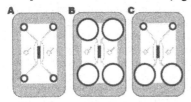

Figure 2: Schematics of different device configurations. A) Four 0.5-mm holes in the inlet ports. B) Four 2.7-mm holes in the inlet ports. C) Two 0.5-mm holes and two 2.7-mm holes.

Device Bonding

Devices were bonded one of two ways depending on the substrate used. For glass microscope slides, the patterned side of PDMS chips and glass microscope slides were exposed to oxygen plasma (30W) for 15 s. Afterwards the PDMS chips were flipped over and placed on top of the exposed microscope slides. The PDMS chips were bonded upon contact to the microscope slides to form operational devices.

For bonding onto microelectrode arrays, a modified version of the "stamp-and-stick" , method developed by Satyanarayana, et al. [3] was used. (Figure 3) First, a small rectangular

58

piece of scotch tape was traced out around the microfluidic chip features and affixed over them to prevent filling of the finer structures in the device. Next, 5 g of uncured PDMS (10:1 by weight) was poured onto a glass microscope slide and spun at 2000 rpm for 5 minutes. The chip was then placed onto the coated glass slide for about 10 minutes and held in place with large binder clips to ensure even coverage. Afterwards, the binder clips were removed and tape pealed off using tweezers. Manual alignment of the chip features with the electrodes was performed under a microscope. With successful alignment, the merged device was placed in a dry oven at 90°C for 15 minutes in order to solidify the bond.

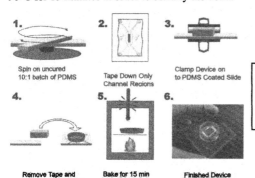

1.	2.	3.
Spin on uncured 10:1 batch of PDMS	Tape Down Only Channel Regions	Clamp Device on to PDMS Coated Slide
4.	5.	6.
Remove Tape and Transfer to Substrate	Bake for 15 min @ 90°C	Finished Device

Figure 3: Schematic detailing the modified stamp-and-stick method used for bonding the microfluidic chip to the microelectrode array.

Device Preparation

Once bonded, devices were sterilized by first exposing them to UV light for 10 minutes. In a biosafety hood 20 μL of 70% ethanol was injected into each chamber and allowed to sit in the hood exposed to UV light for another 30 minutes. Afterwards each chamber was flushed with 40-μL phosphate-buffered saline (PBS) at least twice.

In order to provide a substrate for cells to grow on, poly-L-lysine (PLL, Sigma-Aldrich, St. Louis, MO) solution was used to coat the chamber surfaces. A solution consisting of PLL dissolved in borate buffer at 1 mg/mL was injected into the chambers with a total volume of 20 μL per chamber. 10 mL of sterile water was added into the petri dish to prevent evaporation. The petri dish was taken out of the biosafety hood and allowed to incubate in an incubator (37°C, 5% CO_2) for at least 1 hr.

PC12 Cell Culture

A dividing, neuron-like cell line called PC12 (ATCC, Manassas, VA) was used to evaluate cell viability in the microfluidic devices. PC12 cells were first cultured on surface-treated tissue culture plastic dishes (Primaria, BD Biosciences, San Jose, CA) at a concentration of 10,000 cells/cm^2. PC12 growth media consisted of F-12K Nutrient Mixture (Gibco, Carlsbad, CA) supplemented with 10% fetal horse serum (Gibco), 5% fetal bovine serum(Gibco), penicillin (100 U/mL, Gibco), and streptomycin (100μg/mL, Gibco). PC12 cells can be stimulated to grow neurites by switching to a differentiation media consisting of F-12K Nutrient Mixture supplemented with 50 ng/mL of beta nerve growth factor (R&D Systems, Minneapolis, MN), penicillin (100 U/mL) and streptomycin (100 μg/mL)

After reaching confluence, PC12 cells were passaged and resuspended to a concentration of 3-6000 cells/μL. 20 μL of cell concentrate was added to one chamber of the device. 15 μL of cell concentrate was added to the other chamber of the same device. This established a pressure head differential that caused a small but steady flow between the two different chambers. Media was changed every 48 hr by removing any existing media and reestablishing the 5-μL difference between chambers. Differentiation media was introduced 2-4 days after initial seeding in devices and used for the remainder of the experiment. A sterile water bath of 10 mL was maintained at all times in the petri dishes in order to mitigate evaporation concerns.

Cell Viability Assay

Phase contrast images were taken of the PC12 cells every 24 hr in order monitor daily changes in cell culture in devices. After 8 days, the experiment was ended and viability of remaining cells was qualitatively evaluated with a LIVE/DEAD assay (Invitrogen, Carlsbad, CA). The assay reagents (Calcein AM, 2 mM; Ethidium homodimer, 4 mM) were mixed together in PBS, injected into the device chambers and allowed to incubate for 45 minutes in the dark at room temperature. Afterwards fluorescent microscopy was used to take images of live and dead cells. The images were then merged using the open source software ImageJ.

DISCUSSION

Device validation focused on device assembly and cell viability. Device assembly and disassembly required adapting methods for attaching and removing the microfluidic chip from the MEA. Cell viability work on this device as focused on establishing PC12 cell viability for at least one week. Future work involving neurons that are placed in this device will require time for axons to grow across the axons guidance channels and form connections with the other side. In order to do so, a culture system that can reliably maintain cell viability over extended periods of time is necessary.

Device Assembly

Device assembly and disassembly required protocols be developed for attachment and removal a microfluidic chip from the MEA. Attachment followed a modified stamp-and-stick method described in the Experimental section. The method provided a semi-permanent bond that could provide a tight fluid seal on an uneven MEA surface and strong enough to resist movement during device handling. Since the SAS method was an adhesive type of bonding, it could be reversed with the proper solvent. PDMS tended to swell greatly in the presence of chloroform. [4] Devices soaked in chloroform allowed the disassembly of the chip from the MEA simply by swelling the PDMS chip until in came off the MEA. Any PDMS residue was cleaned off with repeated sonication in chloroform and wiping with a chloroform-soaked cleaning wipe. After removal and cleaning, the MEA was ready to be bonded with another microfluidic chip and used again.

Device Configuration

The challenge of culturing neurons or neuron-model cells like PC12 in microfluidic devices was striking the right balance between nutrient availability versus the shear stress. Due to the low volumes, cells used up available nutrients and accumulated waste faster in their surrounding environments. It thus became necessary to induce some sort of flow through chamber in order to replenish nutrients as well remove waste. However neurons and PC12 cells were sensitive to shear stress that accompanied these media changes and often died due to repeated exposure to excess shear flow. [5]

Extensive testing with the device configuration in Figure 2A reinforced that fact. In this configuration four 0.5-mm holes were punched in the inlet ports of the device. Media changes were facilitated by media-filled glass syringes pushed with syringe pump. However despite adjusting fluid flow rates and times, cultures were never viable for more than 3 days.

Moving to the 2.7-mm hole allowed the formation of media reservoirs by which media could more slowly diffuse into the cell culture chambers. Thus two configurations were tried as seen in Figure 2B and 2C. In Figure 2B, the 0.5 mm holes of Figure 2A are replaced larger 2.7-mm holes. The aim here was to replace the cumbersome syringe pump setup as well as establish a pressure gradient across the axon guidance channels in order to slow down the flow rate to rates found to be acceptable to neurons in literature. [1] However two issues quickly emerged with the Figure 2B configuration. The first was the very swift lateral flow rate caused by the narrow inlet channel connecting the main cell culture chamber to the inlet hole. The second was the unreliable cell seeding that resulted because of the swift flow rate. Cell viability in this device was comparable to Figure 2A.

Figure 2C configuration left the top two holes at 0.5 mm and increased the diameter of the bottom two holes to 2.7-mm holes. This was aimed at retaining the media reservoir and transverse pressure differential driven flow advantages of Figure 2B while slowing the lateral flow rate down enough to maintain cell viability. Initial tests with food coloring showed success in establishing a transverse pressure differential.

Cell Viability

A cell viability assay (Figure 4) was performed on cells grown in the device configuration from Figure 2C. After 8 days, more cells were found alive closer to the 2.7-mm hole (Figure 4, left). After the introduction of differentiation media 48 hrs after seeding, neurites began to extend from the PC12 cells. However upon performing the viability assay, no cells were observed with neurites growing into the guidance channels.

Figure 4: PC12 cells in device configuration from Fig 2C after 8 days *in vitro*. Images of the cell culture chambers are shown in fluorescence (top) and phase contrast (bottom). Staining (top) shows a majority of alive (green) and minority of dead (red) cells. Scale bar is 200 μm.

61

CONCLUSIONS

In order to establish a merged microfluidic/MEA device, several device validation steps were required. Progress toward a fully functional device was demonstrated in device assembly and cell viability. Bonding and removal methods for microfluidic chips from the MEA were tested and proved adequate. Additionally, long term cell viability was demonstrated in the microfluidic devices after exploring different configurations.

Future work will focus on getting neurite in-growth in the axon guidance channels and beginning to culture primary neurons in the device. While there is still much work to be done, the preliminary work is encouraging and thus warrants further investigation.

ACKNOWLEDGMENTS

I would like to thank Amir Shamloo, Dr. Hui Xu, and Dr. Kyle Lampe for helpful discussions in all aspects of the project.

REFERENCES

1. J.W. Park, B. Vahidi, A.M. Taylor, S.W. Rhee, N.L. Jeon. Nature Protocols. 1, 2128 (2006)

2. S Satyanarayana, RN Karnik, A Majumdar. Journal of Microelectromechanical Systems. 14, 392 (2005)

3. GR Prado JD Ross, SP DeWeerth, MC LaPlaca. Journal of Neural Engineering. 2, 148 (2005)

4. JN Lee, C Park ,GM Whitesides. Analytical Chemistry. 23, 6544-6554. (2005)

5. Y Xia and GM Whitesides. Annual Review of Materials Science. 28, 153-184. (1998)

Mater. Res. Soc. Symp. Proc. Vol. 1272 © 2010 Materials Research Society   1272-KK09-07

A New Method of Manufacturing High Speed Aspect Ratio Structures Using SU8 Negative Photoresist

S.K.Persheyev, M.J.Rose

Amorphous Materials Group, Carnegie Laboratory of Physics, Electronics Engineering and Physics Division,
University of Dundee, Scotland, UK

Abstract

SU8 Negative photoresist is finding high demand in applications such as MEMS sensors and waveguides. The possibility of photolithographic patterning and high physical and dielectric properties are attracting ever more users among workers in the electronics industry and increasingly in biomedical applications. In our work we employ an original method of exposing of SU8 and create high aspect ratio structures on glass and other substrates. Dry plasma etching results of negative epoxy-based photoresist by Inductively Coupled Plasma system using gases O_2 and CF_4 are presented.

1. Introduction

The negative epoxy-based SU8 photoresist, due to its special mechanical properties such as durability, water impermeability and dielectric nature upon polymerisation, is often used as a resin for making high aspect ratio, functional MEMS (Micro-Electro-Mechanical Systems)[1-4] device structures and packaging, for cantilevers,[5] and "Lab-on-a-chip"[6] systems. SU8 is ideally suited to the fabrication of devices for micro-fluidic devices,[7, 8] and in bio-MEMS,[2] due its biocompatibility and chemical resistance. The negative photoresist is structured by UV photolithography[3,4,8-10] and structures consisting of multiple layers can be created. The common issues that need to be overcome for conventional through top mask exposure is low adherence for low exposure times and top feature T shape broadening (a phenomenon known as T-topping) for overexposed films. Another limitation is low structure profile flexibility while the same mask is applied.

2. Experimental methods

In our experiments a Karl Suss MJB3 mask aligner was used for UV exposure of the SU8 films whilst soft and hard baking of the films was carried out on a standard hot plate at 95°C. Corning 7059 glass was coated with 120nm of chromium by magnetron sputtering and then patterned with a special mask allowing the formation of 3-5 μm round shaped holes in the metal film. After cleaning the samples were then prebaked at 165 °C in order to remove surface water moisture prior to SU8 deposition. The prebaked samples were immediately spin coated with SU8 and soft baked at 95 °C for 2 minutes, then exposed under the mask aligner upside down without any mask (the patterned chromium acting as a mask as shown on fig.1a,b), with just a clear quartz plate to hold samples. UV exposure was followed by hard (95 °C, 5min) baking for cross-linking of the exposed negative photoresist. For the removal of non cross-linked films after exposure Microposit EC Solvent was applied and then isopropanol (IPA) was used for rinsing cleaning and removal of residual photoresist.

SEM images were obtained by using a JEOL JSM-7400F Field Emission Scanning Electron Microscope with the samples preliminarily coated with thin layer of Au-Pd metal.

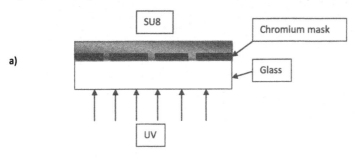

a)

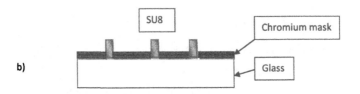

b)

Fig.1 a) Schematics of UV exposure of SU8 negative photoresist, b) SU8 photoresist pillars after developing, removing of non cross-linked material.

One of the advantages of using Corning glass substrates is that they cut off UV radiation below 320nm (fig.2). SU8 is extremely sensitive for deep ultraviolet light which is absorbed mainly near the exposed film surface, and which could compromise structure uniformity.

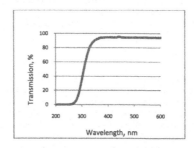

Fig.2. Transmission spectrum of Corning glass

3. Results and discussion

In conventional photolithography of SU8, the negative photoresist will be coated on top of the substrate and then exposed through the masks. It is impossible to change the structure profile by simply changing just the exposure time. The top part of the structure would always see grater irradiation and become more cross linked compared to the lower layer. In the case of underexposure there is a greater possibility that the film will come off the substrate during developing. In this new technique UV exposure is from the substrate side and film structure is now dependent on the exposure dose. Figure 3 shows underexposed photoresist which has sharp needle-like shape (that can of course have their own applications).

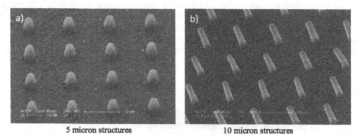

5 micron structures   10 micron structures
Fig.3. Underexposed SU8 negative photoresist

With increasing exposure dose the conical structures become completed cylindrical shapes (fig.4) and it is then a matter of using different viscosity SU8 (SU8 2007, SU8 2025 etc.) to achieve different aspect ratios.

64

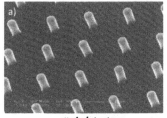

attached structures   harvested structure

Fig.4. High uniformity cylindrical pillar structures (fully exposed)

Further increasing the exposure time will, as expected, enlarge the pillar's top section, making it similar in shape to a light bulb (fig.5). It is clear that with slightly changing exposure angles and high exposure time there is the possibility to make spherical structures, which could be used as light focusing elements for optoelectronic systems.

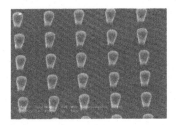

Fig.5. Bulb like shape overexposed SU8 photoresist, top part is larger due to overexposure and reflection from the substrate holder.

Very good mechanical and chemical properties of fully cross-linked SU8 which make it so attractive for MEMS also make it difficult to remove or strip it by conventional methods.[11] Reactive ion etching (RIE) of SU8 photoresist by CF_4 gas showed that etching is not uniform and residual material has a "grassy" structure due polymerisation of photoresist during the etching procedure. The "grassy" structure is clearly seen in figure 6, which also shows the hard texture of the structure walls.

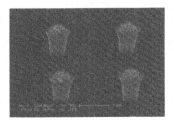

Fig.6. Reactive Ion Etched (RIE) SU8 structures using carbon tetrafluoride (CF_4) gas

Applying oxygen plasma as an etching gas is a well known method for stripping photoresist. Adding a small amount of carbon tetrafluoride substantially increases the SU8 removal rate. Inductively Coupled Plasma (ICP) etching of SU8 negative photoresist allows etching without leaving residual material behind. In fig.7 we demonstrate dry chemical etching of SU8 in a mixture of the gases O_2 and CF_4, where the substrate temperature was kept at 10 degrees Celsius, with a gas pressure of 10mTorr, coil power of 700W and platen power 10W. The highly isotropic etching could be seen under the etch resistive 100nm Al top layer.

65

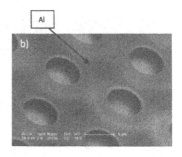

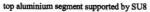

top aluminium segment supported by SU8   etching through holes (aluminium gate electrode)

Fig.7. Inductively coupled plasma etched SU8 structures with aluminium top layer.

4. Future work

The aim of this work was for the first time to demonstrate the possibility of creating high aspect ratio structures using SU8 2007 negative photoresist. In our future research we would like to investigate further practical development of this technique to be able to make various shapes and aspect ratios and the creation of 3D structures. In fig.8 we show preliminary results on fabricating multi-electrode complex devices by ICP etching. Fig.8a shows a single field emitting source in a four terminal configuration, fabricated using negative photoresist and consisting of field emitting material (laser processed amorphous silicon), gate electrode and focusing electrode separated by 1μm and 2μm SU8 photoresist insulator accordingly. Fig.8b is an anchored, but suspended, aluminium structure.

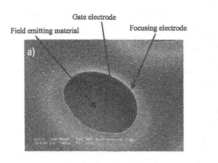

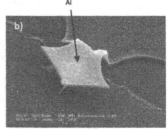

multi-terminal structure   suspended structure

Fig.8. SEM images four terminal of field emitting structure (a) and a free standing aluminium 100nm thick "anchor" (b).

5. Conclusions

For the first time we have demonstrated high aspect ratio SU8 structures achieved by new, inverted substrate UV exposure techniques. We showed that without SU8 etching (wet or dry), and only by photolithographic processing, different- sharp, flat-topped and other profile configurations could be produced. Based on this technique, precise future-size design is feasible. Despite the extra step needed by this method the new process brings reward by adding a greater level of profile flexibility and access to a number of new applications.

6. Acknowledgments

The authors particularly wish to thank Dr. Jose Anguita from University of Surrey, UK for help with the ICP processing of SU8 samples. We are grateful to Dr Gari Harris from University of Dundee for help in obtaining SEM images and fruitful discussions.

References

[1] S.Butefisch, V.Seideman, S.Buttgenbach, Sensors and Actuators A 97-98 (2002) 638-645

[2] Y.Choi, M.McClain, M.LaPlaca, A.Frazier, M.Allen, Biomedical Microdevices, Volume 9, No1, February 2007, pp.7-13(7)

[3] M.B.Chank-Park, J.Zhang, Y.Yan, C.Y.Yue, Sensors and Actuators B 101 (2004) 175-182,

[4] H.Sato, Y.Houshi, T.Otsuka and S.Shoji, Japanese Journal of Applied Physics Vol.43, No. 12, 2004, pp. 8341-8344,

[5] J.H.T. Ransley, M.Watari, D.Sukumaran, R.A.McKendry, A.A.Seshia, Microelectronic Engineering 83 (2006) 1621-1625

[6] J.M.Ruano-Lopez, M.Aguirregabiria, M.Tijero, M.T.Arroyo, J.Elizalde, J.Berganzo, I.Aranburu, F.J.Blanco, K.Mayora, Sensors and Actuators B 114 (2006) 542-551,

[7] Yujun Song, Challa S S R Kumar and Josef Hormes "J.Micromech. Microeng.14(2004) 932-940

[8] B.Bohl, R.Steger, R.Zengerle and P.Koltay, Journal of Micromechanics and Microengineering 15(2005)1125-1130

[9] C.-H.Lee, T.-W.Chang, K.-L.Lee, J.-Y.Lin Appl.Phys.A79, 2027-2031(2004)

[10] K.Elgaid, D.A. McCloy, I.G.Thayne,Microelectronic Engineering 67-68 (2003) 417-421

G.Hong, A.S.Holmes, M.E.Heaton, "DTIP 2003, Mandelieu – La Napoule, France, 5-7 May 2003, pp. 268-271

Mater. Res. Soc. Symp. Proc. Vol. 1272 © 2010 Materials Research Society   1272-KK09-10

Seed assisted phase control of TiOPc: Application of microfluidic mixing

Enkhtuvshin Dorjpalam[1], Hiroyuki Nakamura[1], Kenichi Yamashita[1], Masato Uehara[1], and Hideaki Maeda[1,2,3]

[1] Micro-&Nano Space Chemistry Group, Nanotechnology Research Institute, National Institute of Advanced Industrial Science and Technology (AIST), Tosu, Saga, Japan.
[2] Interdisciplinary Graduate School of Engineering Science, Kyushu University, Kasuga, Fukuoka, Japan
[3] CREST, Japan Science and Technology Agency, Hon-cho, Kawaguchi, Japan

Abstract
Microfluidic mixing was applied to conventional acid pasting process to re-crystallize organic nanocrystals of Titanyl phthalocyanine (TiOPc). TiOPc nanocrystals were re-crystallized in a two step process. Seed particles were prepared by mixing TiOPc, dissolved in concentrated sulfuric acid with deionized water using high speed microchannel mixers. Seed particles were then subjected to post-precipitation treatment to achieve final crystalline product. Effects of seed preparation conditions, such as mixing efficiency (mixer type) and mixing temperature on the structure of final product were studied. Time evolution of optical absorption spectra was examined with a view to elucidate structure evolution during early stages of seed formation process.

Introduction
Recently, photo- and electroactive organic materials have received much research attention [1-3]. In order to realize efficient organic optoelectronic material a precise control over its crystalline structure is essential. In contrast to inorganic materials, which is built of covalent or ionic bonds evenly distributed over the entire structure, organic materials are composed of individual molecules bound together with a weak intermolecular force. Due to this reason, crystal structure tuning of organic materials is very delicate process. In most cases, previously prepared seed material is treated with a various solvents [4-7] to optimize the crystalline structure by means of adjusting the surface energy. In case of this process, reproducibility is most commonly experienced problem. In the present study, we tried to trace effects of seed structure on the structure of final product. The target material is TiOPc, Y-phase of which is recognized as an efficient photoreceptor material [8-9]. Due to the lack of suitable 'good' solvent, solution based preparation techniques except for acid-paste process are not practical for the production of TiOPc. In acid pasting, concentrated sulfuric acid and deionized water are used as the 'good' and 'poor' solvents, respectively. The excessive heat released during initial mixing and the large difference in viscosity of sulfuric acid and water (26.7cP vs. 1.0cP at 20°C) cause inevitable polydispersity from the very beginning of particle nucleation, resulting in non-uniform crystallization. Thus, it is difficult to control the acid pasting process both thermodynamically and kinetically. With a view to overcome this problem, we have adopted microfluidic mixing for seed preparation. Compared to a conventional batch process, better control of heat and mass transfer is expected in microfluidic mixing [10-12].

Experimental

Sample preparation, characterization procedures and main objectives of this study are summed up in Fig.1.

Re-precipitation of the 'Seed' particles: Commercial TiOPc (Aldrich, Dye content: 95%) was dissolved in concentrated sulfuric acid in concentration 1% (w/v) and used as TiOPc precursor. Care was taken to use fresh TiOPc/H_2SO_4 precursor all throughout this study. For a re-precipitation, TiOPc/H_2SO_4 precursor was mixed with deionized water. Cetyltrimethylammonium bromide, CTAB was used as a surfactant in concentration 0.5% (w/w). Mixing was conducted microfluidically using chaos [13] and vortex [14,15] microchannel mixers.

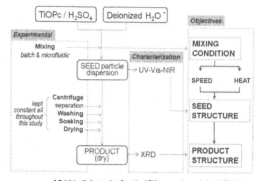

Fig.1. *Sample preparation, characterization procedures and main objectives of this study*

Syringe pump was used to deliver reactants to the mixer. Throughout this study, volume flow rates to feed TiOPc/H_2SO_4 precursor and H_2O were kept constant to 100μl/min and 2300μl/min, respectively. Mixing temperature (25°C, 90°C) was adjusted by using a heater plate, especially designed to mount microchannel mixer on it. In case of low temperature (~4°C) mixing, the mixer was placed on an ice block. For a comparison, re-precipitation was also conducted via batch mixing. In batch experiments, volume ratio was kept same as micromixing (1:23, v/v) and TiOPc/H_2SO_4 precursor was injected at once to a measured amount of deionized water that kept stirring at 1000 rpm. Water (25°C and 90°C) and ice (4°C) bath was used for adjusting mixing temperature. Hereafter we designate as precipitated fresh particles as 'seed'-particles.

Post-precipitation treatment: In this study we adopted post-precipitation or crystallization treatment especially designed to stabilize Y-TiOPc phase. Y-phase has been chosen mostly because of its metastability. Seed particles were separated from the reaction liquid by centrifuge and then washed with deionized water for several times. Washed seed particles were then subjected to crystallization by soaking in H_2O/Toluene mixture under ambient conditions. After soaking for 16 hours, the particles were separated by centrifuge, rinsed with toluene and dried. All throughout this study we kept centrifuge, washing, soaking and drying conditions identical.

Characterization: (1) Structure of seed-particles was examined by optical absorption spectroscopy. A quarts cell with an optical pass length of 0.1 mm was used. Care was taken to keep the aging time constant (time elapsed from the first moment of sample collection to the measurement). (2) Crystal structure of the final product was examined by X-ray diffraction using silicon wafer as a substrate. (3) As a reflection of heat distribution during mixing, surface temperature of mixer is examined by infrared thermal imaging. (4) Time evolution of optical absorption spectra was studied within 10^6ms (10ms – 1h) of time span. Time elapsed from very initial point, when

70

reactants were brought into contact, to the measurement point is considered as a measurement time. For this purpose, 3 kinds of measurement techniques were adopted (Fig. 2). First, in-situ (on-chip) absorption of 10 ms and 22 ms points were measured using microscopic optical absorption technique. UV-VIS spot light source was used. Second, absorption data of 272 ms, 1.3 s, 12 s and 120 s points were measured under flow condition. Measurement time is adjusted by means of fixing tube length. Third, conventional (off-line) optical absorption technique was used to evaluate data of 1h, 24h and 48h points.

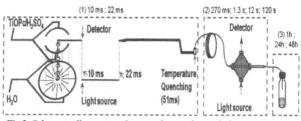

Fig.2. *Schematic illustration of time evolution of optical absorption.*

Results and Discussion

Structure of seed particles were evaluated by optical absorption spectroscopy. Optical and electronic properties of TiOPc strongly depend on its crystal structure and orientation. Besides XRD, optical absorption of TiOPc in UV-VIS-NIR region provides precise and unambiguous information associated with its crystal structure. Optical absorption spectrum of TiOPc is characterized by two main absorption bands, Soret-band in near UV region and Q-band in VIS-NIR region. In particular, Q-band that extends from 550 nm to 900 nm is especially sensitive to the crystal structure. In spite of discrepancies mostly due to the difference in investigated sample, literature data provide definitive information on finger print absorption bands of different crystal phases of TiOPc [16-20]. Optical absorption spectra of seed particles re-precipitated using different mixers at different temperatures are shown in Fig.3. Based on the literature data of structure sensitive absorption bands, Q-band region of each spectrum is fit into four Lorentzian shapes centered at 640, 700, 760 and 840 nm.

Exact position of each spectral component is marked on Fig. 3. Absorption results indicate that seed particles re-precipitated via batch-mixing contain mixture of amorphous (640, 700 nm), phase-I (760nm) and phase-II (840nm) structures. Evolution of phase II structure is more remarkable at low (4°C and 25°C) temperatures. Similarly, mixture of amorphous,

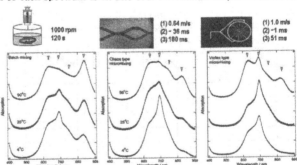

Fig. 3 *Absorption spectra of the seed particles re-precipitated via batch and microfluidic mixing at different temperatures. (1) Linear flow rate, (2) complete mixing time and (3) residence time of micromixers is given. In case of Batch mixing reaction mixture was kept stirring for 2s at 1000 rpm.*

phase-I and phase-II structures was obtained when mixing is conducted using chaos type micro-mixer (linear flow rate 0.64 m/s, complete mixing time 80ms) regardless of mixing temperature. However, it is notable that 760nm spectral component that indicates phase-I structure is most prominent at low mixing temperatures (4^0C and 25^0C). While, when mixing temperature is as high as 90^0C, spectral patterns are almost identical in cases of batch and Chaos type micromixing. On the other hand, when Vortex, the fastest (1.0 m/s linear flow rate, complete mixing time is approx. 1ms) mixer is used, nearly uncontaminated seed particles with amorphous structure was obtained at low (4^0C) temperature. At 25^0C, evolution of Phase-II structure is evident and sample mixed at 90^0C consists of mainly phase-II structure. This result indicates that both mixing pattern and temperature affect structures of seed particles.

Re-precipitation of TiOPc via acid pasting is a highly exothermic reaction, because it is basically a dilution of concentrated H_2SO_4 with water. We tried to evaluate the distribution of mixing temperature by means of heat distribution at the mixer surface. Fig. 4 compares heat distribution at Chaos and Vortex mixer surfaces. Difference in heat distribution at Chaos and Vortex mixer

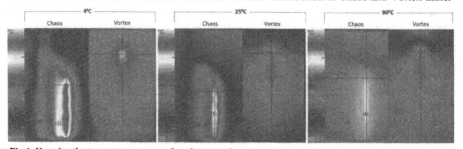

Fig.4. *Heat distribution at micromixer surface during acid pasting process.*

surface is evident. Especially, at lower mixing temperature, Chaos mixer yields much higher maximum temperature, and the high temperature is maintained for long time (large temperature distribution). Chaos mixer was made of glass of which heat conductivity is low. In addition, vo-lume of the reaction mixture that is carried through the mixer is larger due to the complex design of mixing part. Fig. 5 shows XRD patterns of the corresponding final products. In Table 1, crys-tal structure of seed particles and that of corresponding final product is summed up. In most cas-

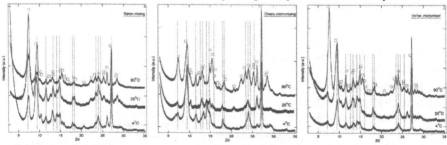

Fig.5. *XRD patterns of crystalline products (dotted line: Y-phase ; circle: Phase-II ; box: Phase-I).*

es, final product is mixture of all Y- , phase-I and phase-II structures. However, when seed particle is prepared using high-speed Vortex mixer at low temperature (4^0C) well defined Y phase resulted. Optical absorption and XRD results clearly indicate that only un-contaminated amorphous seed particles can be transformed into uncontaminated Y-phase product. When seed particle contains phase-I or phase-II components it seems to be difficult to control crystal phases of the final product.

On the other hand, present results indicate that both mixing pattern and mixing temperature affect structure of the seed particles. In our experiments residence time

Table 1. *Crystal phases of the seed particles and corresponding final products*

	Batch		Chaos		Vortex	
	Seed particle	Final product	Seed particle	Final product	Seed particle	Final product
4^0C	Am. + β + α	Y + α + β	Am. + β + α	Y + β	Am.	Ⅺ
25^0C	Am. + β + α	Y + α + β	Am. + β + α	Y + β	Am.+α	Y + β
90^0C	Am. + β + α	Y + α + β	Am. + β + α	Y + α + β	α+Am.	Y + α + β

in other words heating time is very short. It is 180 ms for Chaos and even shorter, 50 ms for Vortex mixer. In other words, in case of Chaos and Vortex mixers, heating is quenched to room temperature after 180 ms and 50 ms, respectively after very initial point of mixing. Therefore, we tried to follow initial seed formation process. For this purpose we examined time evolution of optical absorption spectra within 10^8 ms time span. Vortex mixer was used. The time evolution of optical absorption spectra is shown in Fig.6. Absorbance peaks characteristic of phase-I and phase-II structures increase with mixing temperature, indicating the development of these phases at higher temperature. Even, absorption spectra evaluated at 10 ms point reveal traces of phase-II components and its intensity kept increasing with mixing temperature. This result shows that development of Phase-II and Phase-I structure starts from the very beginning of seed formation

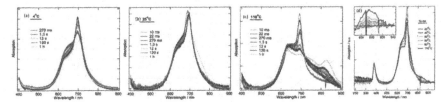

Fig. 6. *Time evolution of the optical absorption spectra of the seed particles prepared using Vortex mixer at different (a) 4^0C, (b) 25^0C and (c) 110^0C. (d) Temperature dependence of absorption spectra at 10ms (780-950 nm region, indicating evolution of phase-II structure is expanded in the inset).*

and preparation conditions of the seed particle affect crystal structure of the final product. Thus it is anticipated that proper control of seed formation process from very beginning may contribute positively to the improvement of process reproducibility.

Summary

Effects of seed preparation conditions on the crystal structure of TiOPc were studied. Amongst various factors, especially distribution of reaction heat evolved during the early stages of seed formation affects crystal structure of final product enormously. Highly efficient mixing at low temperature yielded uncontaminated amorphous seed particles. Upon crystallization treatment, only uncontaminated amorphous seed particles were transformed into Y-phase crystalline product. On the other hand, seed particles that contain even trace phase-I or phase-II components yielded final product consisting of mixed phases. Time evolution of optical absorption spectra proves that development of phase-II structure starts as early as within 10 ms time scale. Thus, in order to obtain TiOPc with a desired crystalline structure it is necessary to control preparation conditions of the seed particles.

References

1) Y. Shirota, J. Mater. Chem., 2000, 10, 1
2) Kock-Yee Law, Chem. Rev., 1993, 93, 449
3) S. A. Jenekhe, S. Yi, Adv. Mater., 2000, 12, 1274
4) M.Knupfer, T.Schwieger, H.Peisert, J.Fink, Phys.Rev.B, 2004, 69, 165210
5) F.Iwatsu, T.Kobayashi, N.Ueda, J. Phys. Chem., 1980, 84, 3223
6) A.M.Hor, R.O.Loutfy, Thin Solid Films, 1983, 106, 291
7) H.R. Chung, E. Kwon, H. Oikawa, H. Kasai and H. Nakanishi, J. Cryst. Growth 2006, 294, 459
8) T. Enokida, R. Hirohashi and T. Nakamura, J. Image. Sci., 1990, 34, 234
9) O.Okada and M.L. Klein, Phys. Chem. Chem. Phys., 2001, 3, 1530
10) S. Krishnadasan, R.J.c. Brown, A.J.deMello and J.C.deMello, Lab on Chip, 2007, 7, 1434
11) K. Mae, Chem. Eng. Sci., 2007, 62, 4842
12) Y. Song, J. Hormes and C.S.S.R. Kumar, Small, 2008, 4, 698
13) A.D.Stroock, S.K.W.Dertinger,A.Ajdari, I.Mezic, GWWhitesides, Science 2002, 295, 647
14) Chung, Y.C., Hsu, Y.L., Jen, C.P., Lu, M.C., Lin, Y.C., Lab Chip, 2004, 4, 70
15) D.A. Waterkamp, M. Heiland, M. Schluter, J.C. Sauvageau, T. Beyerdorff and J. Thoming, Green Chem., 2007, 9, 1084
16) T. Saito, W. Sisk, T. Kobayashi, S. Suzuki, T. Iwayanagi, J. Phys. Chem., 1993, 97, 8026
17) T. Saito, Y. Iwakabe, T. Kobayashi, S. Suzuki, T. Iwayanagi, J. Phys. Chem., 1994, 98, 2726
18) J. Mizuguchi, G. Rihs, H. R. Karfunkel, J. Phys. Chem., 1995, 99, 16217
19) M.Brinkmann, J.C.Wittmann, M.Barthel, M.Hanack, C.Chaumont, Chem.Mater., 2002, 14, 904
20) N. Coppedè, T. Toccoli, A. Pallaoro, F. Siviero, K. Walzer, M. Castriota, E. Cazzanelli, and S. Iannotta, J. Phys. Chem. A 2007, 111, 12550

Mater. Res. Soc. Symp. Proc. Vol. 1272 © 2010 Materials Research Society   1272-KK09-12

Control of the Surface Properties of Microfluidic Devices by Using Polyelectrolyte Multilayer Coatings

Hyun Park,[§] Yoo-Jin An[§] and Sung Yun Yang*

Department of Polymer Science and Engineering, Chungnam National University, 220 Gung-Dong, Yuseong-Gu, Daejeon 305-761, South Korea
§. These authors are equally contributed.

ABSTRACT

Recently, more studies have been conducted in chemical and biological applications using microfluidic or nanofluidic devices.[1] Polymer-based materials have been newly developed in this field due to the great needs of easy processing, cost-effectiveness and clarity for the material. However, it is still challenging to control of the surface properties of these devices on demand. Especially, for biological analysis or detection, micro-fluidics should handle aqueous samples but, most of the current materials in use for micro-fluidic devices are relatively hydrophobic (such as PDMS, PMMA and cyclo-olefin-co polymer, etc). Therefore, they usually need an extra assistance rather than a capillary force to flow the aqueous samples. In this paper, we utilized layer-by-layer deposition of polymer to modify the surface of the micro-channel of the device in order to control surface properties of the micro-channel. We have been studied polyelectrolyte multilayer(PEM) coatings to control surface wettability of the open structures and found various hydrophilic films. Here we demonstrate polyelectrolyte multilayer film as an effective coating for inner surface of micro-fluidic devices to lowering the water contact angle, so that the aqueous fluid will travel smoothly with the channels. Compared to the other surface treatment method such as base cleaning or plasma irradiation, the PEM coating exhibit highly sustained water wettability. Polyelectrolytes used for this study are weak polyelectrolytes including biodegradable polymer such as poly(hyaluronic acid) (HA) for future biological applications.

INTRODUCTION

Microfluidic or nanofluidic devices have been received a great attraction in chemical and biological applications. Due to the increase of the surface/volume ratio, control of the surface properties of these devices is highly important. Polyelectrolyte multilayer films assembled by layer-by-layer are very practical coatings to modify surfaces of various substrates. Layer-by-Layer deposition (LbL) is a versatile technique whereby ultrathin films are assembled from the repetitive, sequential adsorption of oppositely charged polyelectrolytes from dilute aqueous solution.[2-4] Moreover, the functional group which remained reactive after the film deposition allows further chemical reactions such as polymer micro-contact printing, selective photo-crosslinking andcontrolled nanoparticle synthesis.[5-10]

In this paper, we utilized this method to modify the surface of the micro-channel of the device in order to control cellular interactions inside of the channel. We have been studied cellular interactions on various polyelectrolyte multilayer films on open surfaces. Multilayer films comprised of weak polyelectrolytes exhibit different surface properties as they were assembled at different pH conditions. Therefore, the wettability of the PEM-coated also can be varied by depending on the assembly condition. We applied various PEM films on the channel surface of microfluidic device and tested the wetting property and capillary force build-up. Weak polyelectrolytes including biocompatible polymer such as poly(hyalulonic acid) (HA) were investigated for this purpose. And the preliminary results were obtained in the study of surface-cell interaction using these coatings.

EXPERIMENT

Poly(allylamine hydrochloride) (PAH) and poly(acrylic acid) (PAA) were purchased from Polysciences. Hyaluronic acid (HA) was btained from Sigma aldrich. The materials for microfluidic devices in this study are PDMS, PMMA, COC (cyclo-olefin copolymer). Flat substrates such as slide glass as well as the substrates made with the same materials used for micro-fluidic devices were used to obtain the basic properties of the films. These substrates were cleaned by ultrasonic in soap solution. Then the substrates were intensively rinsed with de-ionized water and blown dried with N_2 gas immediately. The coating solutions, polyelectrolyte solutions, were prepared with a concentration of 0.01 M (based on monomer units). The polymer solutions were pH-adjusted to using a 0.01 M HCl or NaOH solution and de-ionized water for rinsing was used without this adjustment. The multilayer film was prepared by a repeated, sequential dipping of the two polymer solutions separated by rinsing steps. The detailed deposition method was followed by the previously published procedure.[8-9] In some cases of micro-channel coating, micro-syringe pump was used to deliver the solution with the lowest pressure.

To study the surface wettability, surface contact angles (static, advancing and receding contact angles) were measured using a contact angle analyzer with pure, deionized water were obtained using the standard sessile drop technique with drops ~2µl in size.

Cellular activity on bare fluidic device and the film-coated device was compared using 293 cells. Cells were cultured on TCPS in Dulbecco's Modified Eagle Medium (DMEM) supplemented with 10 % fetal bovine serum (FBS) at 37°C in a humidified atmosphere of 95% air and 5% CO_2. The pH of the medium was adjusted to 7.4. For attachment and proliferation assays, cells were removed from their growth surface by trypsin and then spun down in a centrifuge at ~1000 rpm for ~5 minutes. The cells were then resuspended in fresh media, mixed in a 1:1 ratio with 0.4% trypan blue (Sigma) and counted with a hemocytometer with trypan blue exclusion to determine cell viability prior to seeding. Cells were seeded onto the sterilized multilayer-coated micro-channels for cell-interaction investigations.

DISCUSSION

The multilayers of poly(allyamine hydrochloride) with poly(acrylic acid) or hyaluronic acid were coated on the inner surfaces of micro fluidic devices. To measure basic properties of the deposited film, the same polyelectrolyte multilayer films were coated on the slides made with the same materials used in the micro-fluidic devices. The polyelectrolyte multilayer films comprised of PAH and PAA or PAH and HA assembled at low pH conditions exhibited high wettability with water. Only a few layers of PAH/PAA or PAH/HA rendered the hydrophobic surface of micro-channels into hydrophilic, generated a capillary force and therefore, the aqueous sample was introduced into the channel without any extra pressure to deliver. Figure 1 shows the dramatic changes in water contact angles of the micro-channeled device. Before having the PAH/PAA coating, the micro-channel created with COC material showed ~110 degrees of water contact angle(Figure 1a) and it was very difficult to put the water droplet into the channel. However, the micro-channel with the polyelectrolyte multilayer-coated exhibit only 30 degrees for the water contact angle and the water dropped in the inlet of the microfluidic device kept traveled all the way through the channel up to the outlet (Figure 1b). In the case of PAH/HA coating, we obtained the very similar results.

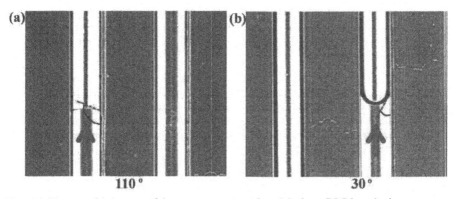

Figure 1. Photographic images of the water contact angles of the bare COC based micro-channels(a) and the PAH/PAA(4.5/3.0) film-coated COC based micro-channels. The red arrows indicate the direction of the water movement in the channel.

To confirm this effect was from the coating, we conducted the controlled experiment of other surface treatment, which can be associated during the micro-fluidic fabrication. Figure 2 shows the water contact angle changes by time after the surface was either coated with the PEM film or treated with plasma cleaning. As shown in Figure 2(a), the water contact angle of the PEM coated surface dropped down to 10 degrees. This initial value slightly increased with time but stabilized at around 30 degrees. In the case of plasma treatment, which is one of the frequent method to treat the surface for hydrophilic character, showed some drop in the water contact angle initially. But, soon the contact angle roused back close to the initial value within a several hours (Figure 2b).

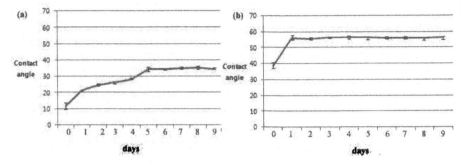

Figure 2. The contact angle changes by time after the surface treatment for micro-fluidic device; (a) PAH/PAA film-coated versus (b) plasma treated samples. (Time of the plasma treatment was also examined in a separate experiment. There was no difference found in the period of plama treatment on contact angle)

Another advantage of this multilayer film would be a good bio-compatibility. We examined the cell viability on these coatings. Cells were introduced in the micro-channels that had the multilayer coating and cultured for a couple of days. We observed good cell viability with these samples and also found excellent block capacity of the films toward the initial non-specific adhesion of cells.

CONCLUSIONS

We have studied the PAH/PAA and PAH/HA polyelectrolyte multilayer films for the micro-fluidic device coating for biosensor application. These multilayers exhibit excellent hydrophilic property with a great deal of stability. They also showed an excellent bio-compatibility toward cells, such as epithelial and blood cells. These films show a great potential as a new functional coating for bio-analytical micro-fluidic devices.

ACKNOWLEDGMENTS

This study was financially supported by the Ministry of Knowledge Economy (MKE) through Korea Evaluation Institute of Industrial Technology (KEIT) grant.

REFERENCES
1. G. M. Whitesides, *Nature*, **442**, 27 (2006).
2. G. Decher, J. D. Hong , *Makromol. Chem. Macromol. Symp.* **16**, 321 (1991)
3. S. Shiratori, M. F. Rubner, *Macromlecules* **33**, 4213 (2000).
4. S. T. Dubas, J. B. Schlenoff, Langmuir **17**, 7725 (2001).
5. S. Y.Yang, , M. F. Rubner, *J. Am. Chem. Soc.*, **124**, 2100(2002).
6. M. C. Berg, S. Y. Yang, P. T. Hammond, M. F. Rubner, *Langmuir,* **20**, 1362 (2004).
7. T. C. Wang, B. Chen, M. F. Rubner, R. E. Cohen, *Langmuir* **17**, 6610 (2001).

8. S. Y. Yang, J. Y. Seo, J. *Colloids and surfaces A : Physicochem Eng. Aspects,* **313-314,** 526 (2008).
9. H.- J. Jeong, W.- H. Pyun, S. Y. Yang, *Macromol. Rapid Comm.* **30(13),** 1109 (2009).
10. S. Y. Yang, D.-Y. Kim, S.-M. Jeong, J. W. Park, *Macromol. Rapid Comm.* **29(9),** 729 (2008).

Mater. Res. Soc. Symp. Proc. VoL 1272 © 2010 Materials Research Society   1272-KK10-03

Fabrication and Applications of Three Dimensional Porous Microwells

Christina L. Randall,[1] Yevgeniy V. Kalinin,[2] Anum Azam,[1] David H. Gracias [2,3]
[1]Department of Biomedical Engineering, [2]Department of Chemical and Biomolecular Engineering and [3]Department of Chemistry, Johns Hopkins University, 3400 N Charles Street Baltimore, MD 21218, USA.

ABSTRACT

In many biological applications, such as cell therapy and drug delivery, there is a need to enhance diffusion by enabling chemical transport in all three dimensions. We highlight this need by comparing diffusion in a conventional two-dimensional (2D) microwell with diffusion in a three-dimensional (3D) cubic microwell using numerical simulations. We also describe the fabrication of hollow polymeric (and biocompatible) cubic microwells and microwell arrays. We emphasize that since the assembly process is compatible with 2D lithographic patterning, porosity can be precisely patterned in all three dimensions. Hence, this platform provides considerable versatility for a variety of applications.

INTRODUCTION

Our concept of a 3D microwell has been described previously [1-2]. As compared to traditional 2D microwells, polyhedra with precisely patterned sidewalls can function as miniaturized analogs, but feature diffusion in all three dimensions. All 3D polyhedral microwells were created using a self-folding strategy that has previously been shown to work across length scales from 100 nm to 2 mm [3-6], with a range of materials including metals, dielectrics and now polymers. The main advantage of having a 3D microwell over a conventional 2D microwell is that interaction of encapsulated contents with the surroundings is enhanced in all three dimensions. Here, we first describe the need for such microwells by comparing diffusion in traditional microwells, which are typically accessible only from one face, and our 3D devices, that feature porosity in all three dimensions. We then describe the creation of polymeric microwells by self-folding. Polymeric microwells are especially advantageous since they are transparent and can be fabricated with biocompatible materials. Finally, we explore the creation of 3D microwell arrays that provide truer analogs of 2D microwell geometries but feature enhanced diffusion in all three dimensions.

METHODS

Numerical Simulations

Three-dimensional numerical models were built in COMSOL Multiphysics 3.5 (COMSOL, Inc., Burlington, MA) and then numerical solutions were sought for the diffusion equation [7]. We assumed that a sample of living tissue 10 μm in size was placed at the geometrical center of the microwell. We also assumed that this tissue consumed all oxygen in its immediate vicinity and that the medium outside of microwells was well mixed. Accordingly, oxygen concentration outside of the microwell was always equal to oxygen concentration in the bulk.

Fabrication of polymeric microwells featuring porosity in all three dimensions

Microwells were fabricated using SU-8 for the panels and poly(ε-caprolactone) (PCL) for the hinges. SU-8 was first spin coated and patterned atop a sacrificial layer on a silicon wafer substrate. Since PCL was not photopatternable using conventional photolithography, we patterned PCL hinges using contact printing which involved fabricating a polydimethylsiloxane (PDMS) stamp and dipcoating the patterned side of the stamp in molten PCL. The stamp was then brought into contact with pre-patterned SU-8 panels on a hot plate set to a temperature above the melting point of PCL (58 °C). This method yielded reliable patterning, but it was challenging to align the master with the samples on the wafer.

We also explored the use of both surface tension and thin film stress-driven folding [8] by evaporating a thin chromium/copper/chromium (Cr/Cu/Cr) trilayer and patterning PCL on top. These polymer microwells self-folded with a thin layer of Cr/Cu adhered to their outer surfaces; the metals were subsequently etched away to form all polymer microwells.

These methods, which yielded some useful results and information about self-folding, had lower yields compared to another method that we tested. In this method, we patterned photoresist hinges on top of patterned SU-8 panels. The substrates were then dip-coated in molten PCL, and excess photoresist and PCL was removed by lift-off dissolution in organic solvents. The 2D panel/hinge structures were then lifted-off by dissolving the sacrificial layer in water, and self-folded into micropolyhedra on heating.

Fabrication of 3D Microwell arrays

Multiple polymers, namely Vetbond (3M), Nexaband (World Precision Instruments) and cross-linked polyurethane (Lutz File & Tool Co.), were investigated as sealing substrates for the microwells. These were typically spin coated at 4000 rpm for one minute on glass slides and tested for their ability to seal the microwells. Additional experiments were then performed to characterize the thickness of the spun sealant to achieve the desired thickness, i.e. approximately 15 μm (to ensure that the sealant covered the lip of the microwell for complete sealing while not covering any pores or decreasing the available volume for loading). This was done by spinning the sealants at increasing spin speeds for one minute. The samples were allowed to rest for an additional minute. Film thicknesses were then measured using a profilometer. Cubic containers with five porous and one open face were fabricated. To make the arrays, individual microwells were manipulated with needles to ensure that the open face was placed down into the spun polymer to form the seal. The microwell arrays were then submerged in PBS and placed in a cell culture incubator to cure.

RESULTS and DISCUSSION

Numerical Simulations

Polyhedra with pores on all faces enhance the interaction of encapsulated contents with the surroundings as compared to more traditional microwells where exchange of chemicals can occur through one open face only. One can expect that, for example, living tissues are going to be better oxygenated when placed within cubes with six open faces. To illustrate this point we present results of numerical simulations (**Fig. 1**) where we have modeled diffusion of dissolved oxygen (D=3×10^{-9} m^2/s) through cell growth medium [9].

In both cases we assumed that a small sample of living tissue (modeled as a 10 μm particle that consumes all oxygen in its immediate vicinity) is placed at the center of a 500 μm microwell that is initially devoid of oxygen (initial oxygen concentration $C_i= 0$). The only way to

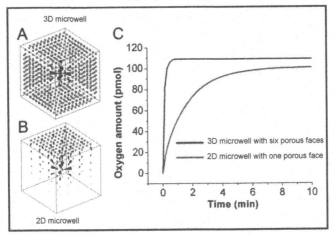

Figure 1. Numerical comparison of oxygen diffusion in a 500 μm 3D microwell with six porous faces as compared to a similarly sized conventional 2D microwell with one porous face. A) and B) present oxygen diffusion fluxes (shown as blue arrows) toward the oxygen sink located at the center of the microwell for a 3D and 2D microwell respectively. The length of the arrows is proportional to the magnitude of the diffusion flux. C) Plot of the total amount of oxygen inside the microwell as a function of time for the case of an abrupt increase of oxygen concentration in the bulk medium.

get oxygen then is via diffusion from the bulk medium (i.e. from the region outside the microwell). It was also assumed that the medium outside of the microwell was well mixed so that there was no oxygen depletion region in the immediate vicinity of the microwell **Fig. 1 A and B** present diffusive fluxes of oxygen toward the sample of living tissue. In the case of the 3D microwell with six open faces, oxygen can diffuse from all six sides as opposed to just one. Accordingly the total oxygen flux in the case of six porous faces is approximately 3.5 times larger than in the case of just one porous face. Microwells with six porous (open) faces also respond faster to changes in the bulk media. **Fig. 1 C** presents the total amount of oxygen that diffuses into the microwell once oxygen concentration increases in the surrounding medium. It can be seen that once oxygen becomes available in the medium (at time t=0) it takes considerably less time to reach a steady state in the microwell with six porous faces as compared to a traditional microwell with only one porous face.

Fabrication of polymeric microwells featuring porosity in all three dimensions

Diffusion-enhancing 3D microwells can be improved for a wider variety of chemical and biological applications by fabricating them from biocompatible polymers rather than metals. There are two crucial challenges associated with transitioning the fabrication process to incorporate this aspect. The polymers chosen must (a) have specific characteristics that enable self-folding and (b) be patternable on the sub-mm scale. Materials used must first pass cytotoxicity tests to prove biocompatibility. Our microwells incorporate square panels that are connected edgewise by hinges, to form the faces of micropolyhedra when self-folded. The material used for the panels must be chemically and structurally stable, optically transparent, and lithographically patternable, to allow for facile pipetting, visibility into the microwell through all of the faces, and precise sidewall porosity, respectively. The material used for the hinges must liquefy at low temperatures to allow for surface tension-driven self-folding, and degrade over a chemically adjustable, biologically relevant time span. Any materials that meet these criteria would function for microwell assembly.

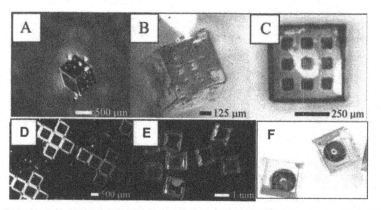

Figure 2. A) Optical image of a porous polymeric microwell after self-folding, prior to Cr/Cu etching. B-C) Top and angled view of A), after wet etching of metal layers. D) SU-8 panels connected by PCL hinges on Si substrate, prior to lift-off and self-folding. E) Optical image of a group of self-folded polymeric microwells showing that the assembly process is parallel. F) Microbeads encapsulated inside polymeric microwells, highlighting the optical transparency of the well.

We were able to fabricate all-polymeric 3D microwells with lithographically patterned pores as small as 8 microns. Advantages of the process include parallel assembly, precise porosity in all dimensions and transparency (**Fig. 2**). Defect modes that were observed in SU-8/PCL polymeric micropolyhedral microwells folded by all methods included missing or mis-adjusted faces, over- and under-folding, and adhered metal content (for methods involving wet etching). Further work would focus on improving yields and utilizing other biomaterials (such as collagen, chitosan) for fabricating the polyhedra for use as vehicles for drug delivery, cell therapy and micro-Petri dishes as well as 3D microwells, since common materials amenable to these applications have been established and their chemical/biological characteristics are well known.

3D Microwell arrays

The concept of a 3D microwell array is shown in **Fig. 3**. Here, polyhedral porous devices are placed on a substrate so that an array is formed which more clearly resembles a 2D microwell plate. The substrate also sealed one face of the container and held it in place. Apart from enabling proper sealing, the choice of the substrate is fairly versatile; here, we explored the use of polymeric substrates. At the present time, polyhedral microwells were positioned on the substrate manually. Eventually, it is envisioned that these could either be self-folded directly on

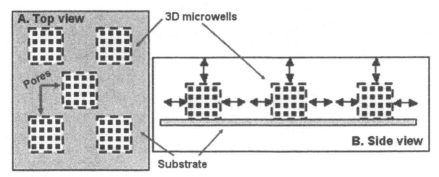

Figure 3. Conceptual schematic of a 3D microwell array. A) Top view and B) side view. The advantage of such a geometric layout is an enhancement in diffusion of encapsulated contents with the surroundings. the substrate to enable fabrication of arrays with large numbers of 3D microwells.

Of the polymer sealants tested, only the cross-linked polyurethane provided a sufficient seal to hold the microwells in place. After completing the spin characterization (**Fig. 4**), we found that the optimum spin speed to produce a cross-linked polyurethane seal of 15 μm was 5000 rpm for one minute. This combination was used to form the microwell arrays.

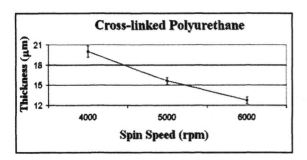

Figure 4. Graph of spin speed versus thickness for the spin-coated cross-linked polyurethane sealant. The thickness was measured using a stylus profilometer. All spin speeds were one minute in duration.

We arranged the individual microwells in a spaced-out pattern to form the arrays (**Fig. 5**). This format allows for the diffusion of potential therapeutic agents (e.g. insulin from beta cells for the treatment of diabetes) as well as the adequate exchange of nutrients and wastes in 3D. Upon curing, we discovered that air bubbles formed around the microwells (**Fig. 5A**). To reduce bubble formation, we modified our array fabrication to be in the curing liquid. Additionally, the microwell arrays were transferred to a desiccator after assembly and a vacuum was applied for five minutes to remove any bubbles that may potentially still form during the cross linking process.

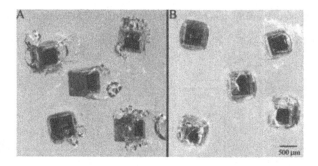

Figure 5. A) 3D microwell arrays after initial curing and B) with reduced bubbles.

CONCLUSIONS

In summary we have described the fabrication and applications of 3D microwells with an emphasis on all-polymeric, biocompatible composition and array construction. Challenges, especially with respect to mass production and parallel loading of cargo, are being addressed to fully enable the creation of these devices for a variety of lab-on-a-chip and biomedical applications. The highlight of the approach is the ability to use conventional 2D lithographic processes to precisely structure porosity in all three dimensions in both individual and arrayed microwells.

ACKNOWLEDGMENTS

We acknowledge funds from the NIH under grant R21 EB007487-02.

REFERENCES

1. T. G. Leong, C. L. Randall, B. R. Benson, A.M. Zarafshar and D. H. Gracias, Lab Chip 8,1621 (2008).
2. D. H. Gracias in Materials and Strategies for Lab-on-a-Chip-Biological Analysis, Cell-Material Interfaces and Fluidic Assembly of Nanostructures, edited by S. Murthy, S. Khan, V. Ugaz and H. Zeringue; Materials Research Symposium Proceedings, Volume 1191, San Francisco, CA , (2009)
3. B. Gimi, T. Leong, Z. Gu, M.Yang, D. Artemov, Z. M. Bhujwalla, D. H. Gracias, Biomed. Microdevices 7, 341 (2005)
4. T. Leong, P. Lester, T. Koh, E. Call, D. H. Gracias, Langmuir 23, 8747 (2007)
5. J. H. Cho, D. H. Gracias, Nanoletters 9, 12, 4049 (2009)
6. T. Leong, A. Zarafshar, D. H. Gracias, Small 6, 7, 792 (2010)
7. E. L. Cussler, Diffusion: Mass Transfer in Fluid Systems, Cambridge University Press, (1997)
8. T. G. Leong, B. R. Benson, E. K. Call, D. H. Gracias, Small 4,1605 (2008)
9. P. Buchwald, Theor. Biol. Med. Model. 6, 13 (2009)

**Directed Assembly and Self Assembly—
From Synthesis to Device Applications**

Mater. Res. Soc. Symp. Proc. Vol. 1272 © 2010 Materials Research Society   1272-LL03-01

Preparation and Characterization of Conjugated Polypseudorotaxanes

Polypyrrole/Cyclodextrins

Yu jie Chen, Wei Wu, Li xing Luan, Wei guang Pu, San xiong He
School of Materals Science and Engineering, East China University of Science and Technology,
130 Meilong Road, Shanghai 200237, China

Abstract:

Conjugated polypseudorotaxanes polypyrrole/α-cyclodextrin and polypyrrole/β-cyclodextrin (Ppy/α-CD and Ppy/β-CD) were synthesized by chemical oxidative polymerization of pyrrole in the presence of α-CD and β-CD respectively, using hydrophobic binding to promote threading of the cyclodextrin units. The FT-IR spectra verified that pyrrole-CD (Py-CD) rotaxanes were presented in the polypyrrole (Ppy) chains. From SEM characterization, neat Ppy and Ppy/β-CD exhibit agglomerated and spherical morphology, while Ppy/α-CD possesses typical dentritic structure with favorable orientation. Based on elemental analysis, the molar ratio of α-CD to pyrrole was 1:5 and β-CD to pyrrole was 1:24. Owing to the irregular Py-CD rotaxanes, the conductivity of Ppy/CDs was decreased by one order of magnitude compared to Ppy. The Ppy/CDs exhibited reduced redox reversibility, as confirmed by an obvious shift of oxidation peak and a lower current in the voltammograms.

Introduction

Cyclodextrins (CDs) are a family of oligosaccharides composed of six or more D-glucopyranose residues attached by α-1, 4-linkages in the $^4C_1$ chair conformation. The most common CDs have six, seven, and eight glupyranose units and referred to as α-, β-, and γ-CD, respectively. Cyclodextrins exhibit a torus-shaped structure with a hydrophobic cavity and a hydrophilic exterior. CDs and their derivatives have been extensively studied as host molecules in supramolecular chemistry [1-2]. Either small molecules or linear polymers can serve as the guests interacting with CDs toward the formation of supramolecular complexes. However, the guest molecules and polymers incorporated into the hydrophobic cavity of CDs must have appropriate size and shape [3-4].

Since its discovery, polypyrrole (Ppy) has been considered an attractive conducting polymer due to good thermal stability, nontoxic, as well as its relatively high electrical conductivity. Storsberg et al. [5] synthesized and characterized the 1:1 host-guest inclusion of cyclodextrin and pyrrole monomer for the first time, then some reports about electro-preparation of polypyrrole in the presence of cyclodextrins appeared [6-7]. Theoretically, all conducting polymers can be applied as molecular wires such as polypyrrole, polyaniline, polyphenylene and polyacetylene. However, resulting conducting products exhibit aggregates structure of powder, particles or film and obviously can't be used as molecular wires. Hence, how to prepare linear π-conjugated oligomers as molecular device becomes an important issue [8].

In this study, conjugated polypseudorotaxanes PPy/CDs has been chemical oxidation polymerized and characterized. The polypseudorotaxane wires composed of α-CD or β-CD as the insulating cover and polypyrrole as the core were expected in this article. It is focused on the morphology and molecular structure of PPy/CDs thus obtained. We investigate and compare the conductivity and voltammetric behavior of pure PPy and polypsedorotaxanes.

Experimental Section

Materials and Characterization

Pyrrole (>97%) was purified by distillation before use. α-Cyclodextrin was obtained from Fluka Chemical Company (Switzerland) and β-cyclodextrin from Sinopharm Chemical Reagent Co., Ltd (China). Other chemicals used were of analytical regent grade without further purification. Distilled water was used in this investigation.

The conductivity of PPy/CDs was measured by the standard four-probe method at 25 ℃. Cyclic voltammetry (CV) experiments were carried out in a standard three-electrode thermostated glass cell at 20 ℃ with a PASTAT2273/CORNERSTONE electrochemical system (Ametek Inc., America). Working electrode was a Pt sheet of 2mm diameter. A Pt foil and a saturated calomel electrode (SCE) were used as the counter and referenced electrodes, respectively. CVs were recorded in 0.1M NaCl. The scan speed was 50mV/s and the potential ranged from -0.5 to 1.8V. FT-IR spectra were measured on a Nicolet 5700 FTIR spectroscopy. The morphology of PPy was examined on a JSM-6360LV scanning electron microscope (JEOL, Japan) and an H-600 transmission electron microscope (Hitachi, Japan). Elemental analysis was performed on a vario EL III element analyzer (GmbH, German).

Chemical oxidative synthesis of PPy/CDs and PPy

The PPy/α-CD was prepared by injecting 2 ml pyrrole to 28.0 g α-CD (Molar ratio of pyrrole: α-CD = 1:1) in 250 ml of distilled water under constant stirring. 57.7 ml of 0.5 M ferric trichloride aqueous solution was added drop-wise into the above mixture under magnetic stirring. The polymerization was continued for about 8 h in 25℃. The resulted solution was filtered and washed by sufficient amount of hot water for one hour to wash out the remaining α-CD and ferric trichloride. The solid product was collected and dried at 60℃ under a vacuum atmosphere for 48 h. The final product was small needle shaped solid. PPy/β-CD was prepared as nearly the same with PPy/α-CD, but the molar ratio of pyrrole: β-CD was 8:1 according to the low solubility of β-CD in water. The resulted product of PPy/β-CD was powder solid. Pure PPy was prepared as above in the absence of any CDs. Finally, the PPy was obtained as platy solid. The physical mixtures of PPy and α-CD, β-CD (PPy+α-CD, PPy+β-CD, respectively) were also made for comparison.

Results and Discussion

Structure Chanacterization and Stoichiometry of the Polypseudorotaxanes

Figure 1 shows the FT-IR spectra of α-CD, β-CD, polypseudorotaxanes PPy/α-CD and PPy/β-CD, and the physical mixtures PPy+α-CD and PPy+β-CD. The spectra of PPy+α-CD and α-CD are nearly the same, except 1543 cm^{-1} suggesting the main characteristic skeleton vibration band of the pyrrole ring. On the spectra of PPy/α-CD, 3364 cm^{-1} (intensive and wide absorption band of O-H stretching vibration), 2926 cm^{-1} (C-H stretching in saturated hydrocarbon), 1154 cm^{-1} (intensive band of ether bond) suggests the existence of α-CD. However, comparing to PPy, the O-H stretching vibration band of PPy/α-CD shifted to low-field. Meanwhile, the corresponding N-H absorption band moved to 3430 cm^{-1} from 3400 cm^{-1} and the band became wider and stronger, which was caused by the hydrogen band between hydroxyl of α-CD and the H of N-H on the backbone PPy. 1089 cm^{-1} in the spectra of PPy is the wavenumber of C-H non-planar bending vibration. But the feeble peak at 1089 cm^{-1} was disappeared in the spectra of PPy/α-CD, which indicates that the PPy/α-CD has the stronger coplanar characteristic of the pyrrole units. That means the α-CD affected the forming of the whole structure of PPy in the chemical oxidative polymerization.

On the FT-IR spectra of PPy/β-CD, the C-H stretching band of saturated hydrocarbon weakly appears at near 2918 cm^{-1}, furthermore, the C-H non-planar bending vibration band of backbone PPy still exists at near 1089 cm^{-1}. The spectrum of PPy/β-CD is nearly the same with PPy+β-CD, but the absorption peaks of β-CD are more intensive in the spectra of PPy+β-CD. These results indicate that PPy/β-CD has thimbleful β-CD, moreover, β-CD has less influence on backbone PPy than α-CD because of the too low content of β-CD in the reactants.

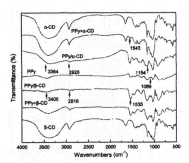

Figure 1 FT-IR spectra of α-CD, PPy+α-CD, PPy/α-CD, PPy, PPy/β-CD, PPy+β-CD, and β-CD.

Elemental analysis was performed on PPy to determine the amount of α-CD encapsulating PPy. The sample was dried at 80⊠ in the vacuum oven for another 72h to remove the residual water completely. The PPy/α-CD sample has a [O]/[N] value of about 6.69 and a [C]/[N] value of about 10.35 (see tab.1), then the molar ratio of α-CD to pyrrole is about 1:5. This result indicates that each five pyrrole molecules have one rotaxane structure with one α-CD molecule. Meanwhile, the PPy/β-CD sample has a [O] /[N] value of about 1.18 and a [C]/[N] value of about 4.29, then the molar ratio of β-CD to pyrrole is about 1:24. This result indicates that each twenty-four pyrrole molecules have one rotaxane structure with one β-CD molecule, which is far away from the expected result 1: 1.

Because the pyrrole-radical excluded from the CD cavity, which impeded the formation of the Py-CD rotaxane and degraded the speed of chain growth, not every pyrrole unit in the chain became rotaxane with CD. The molar ratio of β-CD reactant to pyrrole reactant was 1:8, α-CD to pyrrole was 1:1, therefore, it can be concluded that about 33% β-CD molecules successfully encapsulated PPy chain and only 20% α-CD molecules encapsulated PPy. It demonstrates that the formation of polyrotaxane structure is influenced by the size of cyclodextrin cavity, which means the cavity is larger, the formation of polyrotaxanes is easier.

Table 1 The result of elemental analysis

Sample	N %	C %	H %	O %
PPy/α-CD	5.37	47.65	5.93	41.05
PPy/β-CD	15.88	58.42	4.27	21.43

Morphology of PPy/CDs and PPy

Figure 2 shows the SEM spectra of PPy, PPy/α-CD and PPy/β-CD. PPy prepared in the absence of CDs resulted in a spherical morphology as shown in Figure 2a, because α-position and β-position of pyrrole monomer have the same polymerization capacity, which easily induces crosslinking to agglomerated and spherical PPy. There are more rod-like and fiber-like substances in the bulk of PPy/α-CD (Figure 2b) than PPy. One typical dentritic structure of PPy/α-CD is enlarged in Figure 2c and displays favorable orientation and organization. It indicates that the α-CD molecules might slide along and rotate around the polymer chains, and some α-CD molecules formed the pseudopolyrotaxanes with the polymer axes. However, PPy/β-CD forms flocculent aggregates (Figure 2d) and shows little orientation and organization, which is caused by too low content of β-CD in the polymerization.

The good organization of PPy/α-CD and agglomerate of PPy are demonstrated evidently as shown in Figure 3.   We can conclude the interaction of α-CD and polypyrrole could strengthen the binding force between polymer chains and finally optimize the configuration of polymer. This result is in accordance with the electrochemical polymerization which was reported by Izaoumen and coworkers[6]. Figure 3a shows that PPy/α-CD was fibrous with a diameter of about 250 nm, which indicates that the PPy/α-CD is potential in application as molecular wire.

(a) pure PPy   (b) PPy/α-CD

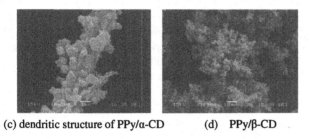

(c) dendritic structure of PPy/α-CD   (d)   PPy/β-CD

Figure 2 SEM spectra of pure PPy (a); PPy/α-CD (b); dendritic structure of PPy/α-CD (c); PPy/β-CD (d).

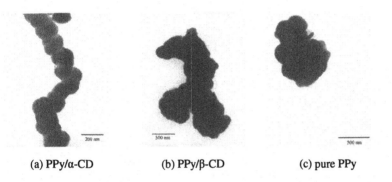

(a) PPy/α-CD   (b) PPy/β-CD   (c) pure PPy

Figure 3 TEM spectra of PPy/α-CD (a) ; PPy/β-CD (b);   pure PPy (c).

Conductivity and Voltammetric Behavior of Polypseudorotaxanes

Table 2 shows the conductivity of PPy/CDs at room temperature is decreased by one order of magnitude as compared with pure PPy. This result implies that CDs are totally or partly thread into the chain of PPy macromolecule, not only act as dopants in the conductive polymer, or else the conductivity wouldn't decrease so significantly. The presence of CDs in PPy/CD weak the interchain charges transition, because CDs and PPy form the polypseuorotaxanes by electrostatic force or hydrogen bond. This also can be explained by stereo-hindrance effect.

Table 2   The results of conducting test

Sample	PPy	PPy/α-CD	PPy/β-CD
Conductivity/S·cm^{-1}	5.38	0.3694	0.4537

The voltammograms of PPy and PPy/CDs were carried out in 0.2M NaCl, and CVs after the 30th cycles are presented in Figure 4. Figure 4a shows a first strong oxidation peak around 1.46V. Nevertheless, the fist oxidation peak of PPy/α-CD shifts to 1.32V and the current of the oxidation peak was apparently weaker than PPy. The slight negative shift of the oxidation peak and the lower current compare to PPy might be the results of the polypyrrole partly or entirely include in the α-CD hydrophobic cavity. The first oxidation peak of PPy/β-CD closes to PPy, but the current is medium comparing with PPy and PPy/α-CD. The number of α-CD molecules threaded by PPy (molar ratio 1:5) is much larger than the number of β-CD (molar ratio 1:24), hence the oxidation peak shift of PPy/α-CD is quite obvious, but PPy/β-CD has no evident oxidation peak shift.

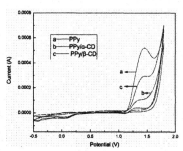

Figure 4 Cyclic voltammograms of (a) PPy; (b) PPy/α-CD; (c)PPy/β-CD; electrolytic solution: 0.2 M NaCl; sweep rate: 50mV/s.

Conclusions

The conjugated polypseudorotaxanes PPy/α-CD and PPy/β-CD are synthesized by oxidative polymerization of pyrrole in the presence of α-CD and β-CD respectively. The molar ratio of α-CD to pyrrole monomer is 1:5 and β-CD to pyrrole is 1:24. Compared to polypyrrole, Polypseudorotaxanes have obviously decreased conductivity and reduced redox reversibility in terms of the discontinuous presence of Py-CD rotaxanes in the PPy chain. However, TEM shows some polypseudorotaxanes PPy/α-CD were fibrous with a diameter of about 250 nm, which could promise its potential in application as molecular wire.

References

1. R. Villalonga, R. Cao, A. Fragoso, *Chem. Rev.* **107**, 3088-3116 (2007).

2. F. Hapiot, S. Tilloy, E. Monflier, *Chem. Rev.* **106**, 767-781 (2006).

3. G. Wenz, *Angew. Chem. Int. Ed. Engl.* **33**, 803-822 (1994).

4. C. Torque, H. Bricout, F. Hapiot, E. Monflier, *Tetrahedron* **60**, 6487-6493 (2004).

5. J. Storsberg, H. Ritter, H. Pielartzik, L. Groenendaal, *Adv. Mater.* **12**, 567-569 (2000).

6. N. Izaoumen, D. Bouchta, H. Zejli, M. E. Kaoutit, A. M. Stalcup, K. R. Temsamani, *Talanta* **66**, 111-117 (2005).

7. N. Izaoumen, D. Bouchta, H. Zejli, M. E. Kaoutit, K. R. Temsamani, *Analytical Letters* **38**, 1869-1885 (2005).

8. J. M. Tour, *Chem. Rev.* **96**, 537-553 (1996)

Mater. Res. Soc. Symp. Proc. Vol. 1272 © 2010 Materials Research Society 1272-LL03-10

A Facile, One-Pot Synthesis of Polyaniline-Au Ultralong Nanowires and Electric-Field-Directed Assembly toward Organic Electronic Devices

Nam-Jung Kim*[1], Jian Jiao[2] and Jae Wan Kwon[2]
[1]Mechanical and Aerospace Engineering, Univ. of Missouri, Columbia, MO 65211, U.S.A.
[2]Electrical and Computer Engineering, Univ. of Missouri, Columbia, MO 65211, U.S.A.

*Corresponding Author: Nam-Jung Kim, Ph. D., kimna@missouri.edu.

ABSTRACT

Polyaniline-Au composite nanowires (NWs) and ribbons were synthesized via a charge-exchange redox process in a bi-phase interfacial polymerization method. The spontaneously-nucleated metal-polymer nanoparticles grow further to constitute Au-rich NWs via oriented self-assembly, and parallel assembly lines of NWs form a microns-wide, millimeters-long composite ribbon showing an excellent structural ordering and uniformity over the long NW axis. The structural properties of the composite nanostructures were investigated by SEM, TEM, and AFM. We also found that the resulting NWs/ribbons have a great potential for uses in hybrid electronic devices that are easily scalable and integrated. The charge transfer/trapping models can be applied to explain the experimental conductance data from the polymer-metal NW devices.

INTRODUCTION

We present a novel self-assembly phenomenon to form ultralong, uniform polyaniline (PANI)-Au composite NWs and ribbons. PANI nanostructures have been drawing huge attention due to their unique physical/chemical properties and a variety of technological uses in flexible nano electronics, memory devices, photovoltaic cells, and conductive coatings.[1] While pure PANI films or nanofibers produced by chemical or electrochemical oxidation process have been intensively studied,[2, 3] more interesting results can be explored by incorporating metallic nanoparticles with the polymer materials.[4-6] However, most of the prior attempts could produce relatively short (about only a few µm) 1-D nanotubes or fibers without a long-range ordering.

In contrast, we have recently produced highly-ordered, ultralong PANI-Au NWs/ribbons through an anisotropic self-assembly growth during a bi-phase polymerization process. Here we first report macroscopically-uniform composite nanostructures made of conducting polymers as a result of synergistic effects from Au interacting with PANI. Electron microscope images reveal the self-assembled Au NWs and the formation of micrometers-wide PANI ribbons decorated with parallel lines of the Au NWs. AFM analysis provides the detailed information of various NW diameters and the nanoscale roughness of PANI ribbons. The underlying assembly mechanisms of the ordered nanostructures are discussed. Electrical properties of composite NWs/ribbons are also studied by using microelectrode-patterned Si substrates. Interesting DC *I-V* characteristics are found, and a charge-trapping model is applied to understand the electric-field-dependent transport behaviors in the composite micro channels.

EXPERIMENTAL DETAILS

All the chemicals (aniline and Au (III) chloride solution) were purchased from the Sigma-Aldrich Chemical (St. Louis, MO, USA). The bi-phase solution was prepared by adding the aniline (2 ml) to the diluted aqueous Au solution (10 ml) at room temperature. UV-visible spectroscopy (UV-1600, Shimadzu) was used to monitor the AuNP formation and PANI polymerization process at different times. FESEM (Quanta FEG 600 ESEM, OR, USA) and AFM (Hysitron Inc., MN, USA) was used to characterize the PANI-Au nanostructures. Micropatterned Au electrodes were prepared by a standard optical lithography and were used to assemble the NWs/ribbons across the electrodes and measure the *I-V* characteristics.

RESULTS and DISCUSSION

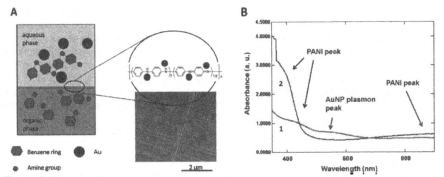

Fig. 1. (A) Schematic diagram of an organic-water bi-phase system and proposed formation of PANI-Au nanocomposites near the interface. The inset shows a typical SEM image of highly-ordered PANI ribbons-Au NWs deposited on a conducting substrate. Parallel Au-rich NWs appear bright in the PANI ribbon. (B) UV-visible absorption spectra at early (1) and later (2) stages of PANI polymerization, respectively. PANI polymer peaks increase with the increased reaction times in solution.

Fig. 1A illustrates the synthesis process of PANI-Au composite nanostructures in a bi-phase water-aniline interface system. Through the well-sustained interface, aniline monomers interact with Au ions dissolved in the water phase via a charge exchange redox process. In early stage of redox process, immediate AuNP formation can be inferred by observing the water solution to change the color from pale yellow to wine red. Then the water phase gradually appears to be dark red and greenish, indicating the formation of polymerized nano composites. This NP nucleation and polymerization process is further elucidated by the UV-visible spectroscopy acquired at different times in Fig. 1B. Initially the AuNP surface Plasmon peak at 530 nm is evident. As the reaction proceeds, the Plasmon peak noticeably decreases, while PANI polymer characteristic peaks are increasing.[7] It is well known that the appearance of AuNP Plasmon is sensitive to the nature of surface capping layers or the dielectric properties of colloidal suspension. With a prolonged reaction time, the polymer-capped AuNPs can further grow to larger 1-D nanostructures due to an anisotropic template effect. However, UV-visible absorption data from the bulk sample is not sensitive enough to display the onset of Au NW and PANI ribbon formation during the course of self-assembly growth. Thus, we examined colloidal composite nanostructures using electron microscopes by collecting sample droplets and depositing them on a conducting substrate for a high-resolution image analysis.

Fig. 2 presents SEM micrographs of PANI ribbons/Au NWs drop-deposited on a Au-coated glass slide as a conducting substrate. It is evident that long lines of Au-rich NWs are orderly assembled in a uniform ribbon that can be as long as millimeters and as wide as micrometers. Multiple lines of Au NWs appear brighter in the SEM image presumably due to an enhanced height of NWs that is further analyzed by AFM scans. Energy-dispersive X-ray spectroscopy (EDS) data also confirms that the NW lines are rich in Au elements. The ribbon folding, bending and splitting are consistently observed, which is characteristic of the flexible polymer ribbon. Some multiple-layered ribbons are found to be easily peeled apart as expected in graphene-like structures,[8] and each separate layer of the ribbon shows an identical nanostructure with the other layers, indicating the possibility of acquiring monolayer PANI ribbons of the same morphology. Besides the structural uniformity demonstrated by the SEM micrographs, micro-Raman studies confirm the presence of the polymer layer and the chemical/molecular uniformity of PANI-Au composite NWs along the axis. From the micro-Raman results, it is suspected that Au ions/atoms not only oxidize the aniline monomers through a charge exchange but also attach to developing PANI products in a site-specific manner via Au-N binding (Raman experimental data will be published elsewhere).

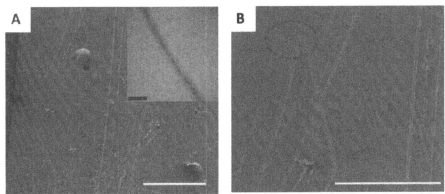

Fig. 2. (A) SEM micrographs of PANI ribbons/Au NWs deposited a solid substrate. The inset in (A) provides a TEM image of a Au-rich NW (scale bar is 50 nm). (B) The ribbon folding (marked by a circle) and bending (denoted by an arrow) are evident. Scale bars are (A) 50 and (B) 30 μm, respectively.

AFM analysis reveals the typical thickness of ribbons and diameters/heights of embedded NWs (Fig. 3). Each Au NW parallel to a long axis of the ribbon shows a superb uniformity in diameter over a long distance, while the diameters of each Au NW can vary in the range from a few nm to 100 nm. The NW diameter of ~10 nm can be hundreds of microns long. So the aspect ratio of a typical NW is estimated to be 10^4. The flat surface of PANI ribbon is found to be extremely smooth. Interestingly, from more AFM images, thick NWs (d > 100 nm) frequently show the linearly-aligned rod-like structure, while nanoscale gaps are apparent between neighboring nano rods. Those fat nano rods seem to orient well along the sharp edges of the present PANI ribbons, while they are drop-deposited onto a Au-coated glass substrate and dried in air before the microscope image acquisition.

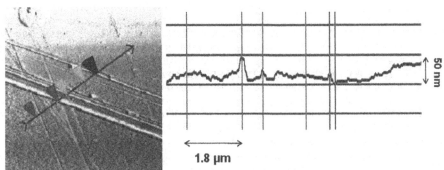

1.8 µm

Fig. 3. AFM topological image (10 µm in size) and height profile along the scan line marked in the image.

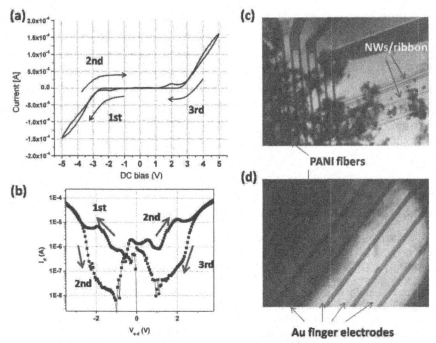

Fig. 4. (a) *I-V* characteristics from NWs/ribbon aligned across Au electrodes. (b) The same *I-V* data with (a) in a log scale of the current axis. (c) Optical image of PANI ribbons aligned on the electrodes. (d) Optical image of PANI nanofiber bundles produced electrochemically near the AC biased electrode.

Underlying growth mechanisms of PANI-Au nanostructures are of great interest. There may be three cooperative growth mechanisms proposed to explain the 1-D ordered assembly of PANI-Au composites: 1) PANI itself has a linear backbone structure composed of repeated aniline oligomers so that PANI chain can grow predominantly in the long-axis direction; 2) AuNPs can contain themselves inside the PANI shells and grow anisotropically through polymer capping effect; and 3) a side-by-side parallel assembly of long PANI chains forms a ribbon structure in the solution or on a solid substrate, and the ribbon may host the colloidal AuNPs/NWs in an orderly-oriented way. Therefore, the combined effects of AuNP attachment to the polymer and the linear backbone of PANI chains would result in an unprecedented macroscopic-long, highly-oriented polymer-metal nanostructure.

The electric-field-induced or dielectrophoresis (DEP) assembly of NWs/ribbon across Au electrodes was investigated, while a set of multiple electrodes was designed and an AC voltage bias (4 V, 100 Hz) condition was used (Fig. 4). The immediate advantages of using a series of Au electrodes patterned on Si substrate are the chemical inertness of Au and different electrode pairs available for AC bias contacts for DEP-directed assembly and DC bias contacts for the conductance measurements. Under the DEP condition, aligned PANI ribbon fully stretches from the biased electrode surface to the grounded electrode area that is viewed in Fig. 4c. During the directed assembly, PANI nanofiber bundles were also produced due to the electrochemical polymerization near the AC biased Au electrode surface. Thus, the DC I-V data was obtained by choosing a middle finger electrode pair in order to avoid the interference from the nanofibers. Similar to the prior results in the literature, low (OFF) and high (ON) current states exist and those currents differ by two orders of magnitude at 2 or -2 V.[6, 9] A negative differential resistance (NDR) and an interesting electrical hysteresis behavior are observed in a certain range of DC bias. On the other hand, I-V curves appear very symmetric on both sides of DC biases. We attribute the remarkable DC bias symmetry to a uniform physical contact between PANI ribbon and different Au electrodes used for the conductance measurement. The I-V data was found to be highly reproducible and stable over many test cycles. All fabrication and measurement was conducted at room temperature in the ambient lab environment.

The complex nonlinear I-V data can be qualitatively understood by the charge injection and trapping model. In the first scan (Fig. 4), the NW embedded ribbon is initially in a highly conducting state so that a sharp rise of the current is manifest near the zero bias. Then the charge injection from Au electrode should govern the current. But when the DC bias approaches -2 V, the charge begins to be trapped in AuNPs embedded in PANI. Thus, the current decreases with the increasing bias (entering NDR region). After the charges are fully trapped at the NP interface, the current starts to increase again while entering the space-charge-limited regime. Note that the trapped charges may remain at the maximum state as long as the applied field is above a threshold voltage (-1 V in this case). In fact, the current starts to decrease rapidly around -2 V in the second return scan in the presence of fully charged NPs as impurity sites until the bias reaches -1 V at which the stored charge becomes unstable and a rapid discharge takes place. After all the charges are released, the conducting channel returns back to the high-current state. The same charge trapping and release process can be applied to the other side of DC bias range. The presence of metal NP is known to be crucial to observe the conductance hysteresis from polymer channels.[10] Further experiment is needed to fully understand the charged states of NPs.

99

CONCLUSIONS

Highly-ordered PANI-Au composite nanostructures were synthesized and characterized by various tools. The cooperative interaction between polymers and metallic NPs during the self-assembly growth seems important to fabricate such a well-organized nanostructure that is physically and chemically uniform in a macroscopic scale. Au NW embedded PANI ribbons are mechanically flexible and multi-layered structures.

PANI composite NWs/ribbons can align themselves to cross the electrode gaps by simply drop-depositing the colloidal sample solution on the micropatterned substrate under the presence of electric field. The obtained nonlinear I-V curves strongly indicate the interface charge trapping effects on the DC transport that varies by orders of magnitude, depending on the applied bias conditions. This work combining bottom-up and top-down approaches provides a promising example how nanoscopic materials can be effectively integrated into existing microscopic patterns and hence the usefulness of nanomaterials can be significantly increased.

ACKNOWLEDGMENTS

The authors gratefully acknowledge the NW conductance measurements by Tae Kyeong Kim in Dr. Seunghun Hong's group in School of Physics, Seoul National University, Korea. Nam-Jung Kim also thanks Prof. Khanna at the Univ. of Missouri for his generous help with AFM image acquisition.

REFERENCES

1. D. Li, J. Huang and R. B. Kaner, *Accounts of Chemical Research*, 2009, **42**, 135-145.
2. Huang J and Kaner R, *Journal of the American Chemical Society*, 2004, **126**, 851-855.
3. L. Y. O. Yang, C. Chang, S. Liu, C. Wu and S. L. Yau, *Journal of the American Chemical Society*, 2007, **129**, 8076-8077.
4. J. M. Kinyanjui, D. W. Hatchett, J. A. Smith and M. Josowicz, *Chemistry of Materials*, 2004, **16**, 3390-3398.
5. A. Drury, S. Chaure, M. Kroll, V. Nocolosi, N. Chaure and W. J. Blau, *Chemistry of Materials*, 2007, **19**, 4252-4258.
6. R. J. Tseng, J. Huang, J. Ouyang, R. B. Kaner and Y. Yang, *Nano Letters*, 2005, **5**, 1077-1080.
7. J. Huang, S. Virji, B. H. Weiler and R. B. Kaner, *Journal of the American Chemical Society*, 2002, **125**, 314-315.
8. A. K. Geim and K. S. Novoselov, *Nature Materials*, 2007, **6**, 183-191.
9. R. J. Tseng, C. O. Baker, B. Shedd, J. Huang, R. B. Kaner, J. Ouyang and Y. Yang, *Applied Physics Letters*, 2007, **90**, 053101.
10. L. D. Bozano, B. W. Kean, M. Beinhoff, K. R. Carter, P. M. Rice and J. C. Scott, *Advanced Functional Materials*, 2005, **15**, 1933-1939.

Mater. Res. Soc. Symp. Proc. Vol. 1272 © 2010 Materials Research Society   1272-LL09-02

A Combinational Effect of the Activated Alloy Phase Separation and Strain Relaxation in the Improved Thermal Stability of Coupled Multilayer S-K Grown InAs/GaAs Quantum Dots

S. Adhikary, K. Ghosh and S. Chakrabarti

Center for Nanoelectronics, Department of Electrical Engineering
Indian Institute of Technology Bombay, Mumbai - 400076, Maharashtra, India.
Corresponding author email: subho@ee.iitb.ac.in

Abstract

Effect of post growth annealing on 10 layer stacked InAs/GaAs quantum dots with InAlGaAs/GaAs combination capping layer grown by molecular beam epitaxy has been investigated. The quantum dot heterostructure shows a low temperature (8 K) photoluminescence emission peak at 0.98 eV. *No frequency shift in the peak emission wavelength is seen even for annealing up to 700°C which is desirable for laser devices requiring strict tolerances on operating wavelength.* This is attributed to the simultaneous effect of the strain field, propagating from the seed layer to the active layer of the multilayer quantum dot and the activation alloy phase separation effect that is driving indium atoms towards the periphery of the quantum dots due to the presence of a quaternary InAlGaAs layer. Higher activation energy (of the order of ~ 250 meV) even at 650°C annealing temperature also signifies the stronger carrier confinement potential of the quantum dots. All these results demonstrate higher thermal stability of the emission peak of the devices using this heterostructure.

Introduction

Self-assembled InAs/GaAs quantum dots (QDs) are the potential candidates for superior performance lasers [1,2] and detectors [3,4] due to their unique δ like density of states [5], low threshold current [6] and high characteristic temperature [7]. Multilayer QD (MQD) heterostructures are widely employed to increase the modal gain of the QD lasers which also provides higher carrier localization, increasing the efficiency [8-10]. High temperature annealing is required for the growth of good quality cladding layers in long-wavelength lasers. However, this leads to pronounced blue-shifting of the ground state (GS) photoluminescence (PL) emission from QDs due to change in the composition of the dots caused by In/Ga interdiffusion. Along with this, the formation of defects in high temperature annealed QDs leads to a decrease in PL intensity [11, 12].

In this article, we have focused on the annealing of multilayer (ten layer stacked) InAs/GaAs QDs with a combination capping layer of GaAs (180 Å) and quaternary InAlGaAs (30Å) which does not exhibit a blue-shift in the PL emission peak even for high annealing temperatures upto 700°C. This is desirable for the fabrication of devices which require strict tolerance on the emission wavelength such as distributed feedback lasers and vertical cavity surface emitting lasers (VCSELs). We have also investigated the thermal quenching of integrated PL intensity of the as-grown and annealed samples to ascertain the dislocations generated in the QDs and the carrier confinement potential.

Experimental method

Ten layer stacks of InAs QDs on a semi insulating (001) GaAs substrate was grown by solid source molecular beam epitaxy (MBE). First an intrinsic GaAs buffer layer of thickness 0.4 μm was grown at 590°C. Then the substrate temperature was gradually reduced down to 520°C with the growth of 1000Å intrinsic GaAs. Following this InAs QDs were grown at a rate of 0.2 ML/sec at 520°C. Each dot layer is separated by combination of quaternary $In_{0.21}Al_{0.21}Ga_{0.58}As$ layer of thickness 30Å and GaAs layer of 180Å grown at 520^0C .The details of the heterostructure were published elsewhere [13]. The grown samples cut from the central region of wafer into pieces were subjected to *ex situ* rapid thermal annealing (RTA) in AnnealSys AS-One 150 RTP system for duration of 30s at different annealing temperatures such as 650°C, 700°C, 750°C, 800°C and 850°C in an argon atmosphere. Temperature dependent PL experiment has been carried out in the temperature range between 8K and 305K with the sample mounted in a continuous flow He-cryostat using 405nm excitation wavelength under 40mW power. The signal was dispersed by a 0.75nm monochromator and detected by liquid nitrogen cooled InGaAs array detectors.

Results and discussion

The low temperature (8K) PL spectrum for as-grown ans as well as annealed samples is shown in figure 1. The ground state PL emission peak is observed at 0.98eV which is red-shifted as compared to the emission peak of a single layer InAs/GaAs QD system [14], implying the formation of large sized dots. In reference [14] the single layer QDs are grown with similar monolayer coverage (2.7 ML) and growth rate (0.2 ML/s) as that of the MQD sample investigated in the present research. The larger sized dots in the active layer of the MQD sample is due to the activation alloy phase separation (AAPS) effect of InAlGaAs quaternary alloy driving the indium atoms at the periphery of the dots [15] and the strain field propagation from seed layer to the upper layers of the heterostructure known as the templating effect which helps in the nucleation of the upper QD layers [13, 16-17]. A second PL peak is also seen at a shorter wavelength in figure 1. The separation between the two peaks is about 100 meV, indicating that the two peaks are due to the formation of two families of QDs with different sizes.

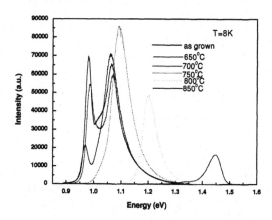

Figure 1. PL spectra for as-grown and annealed multilayer InAs/GaAs QDs.

Since high temperature annealing causes the strain enhanced interdiffusion of indium from the QDs into the barrier and of gallium in the reverse direction [18, 19], PL emission of the annealed QDs is expected to shift at a shorter wavelength. However, no significant shift in the emission peak is seen upto annealing temperatures of 700°C in the MQD sample under investigation which is possibly due to two reasons. Firstly, by the suppression of In/Ga interdiffusion by the presence of In rich region in the periphery of the QDs due to the AAPS effect as explained [15]. The second effect is the strain field propagation from the seed layer to the upper layers of QDs, which not only causes the islands to be vertically stacked, but also maintains a strain relaxed state in the QD system [13]. Both the samples annealed at 650°C and at 700°C have two PL peaks, which like the as-grown sample is possibly due to presence of two families of dots. Interestingly, it is observed from the PL spectra that with increase in annealing temperature, the intensity of the peak at longer wavelength due to large size dots is significantly reduced, while the one at shorter wavelength side due to the smaller sized QDs is enhanced. Redistribution of indium atoms due to thermal annealing which causes dissolution of the larger dots and causing a more uniform dot size density can be ascribed for this phenomenon. As the annealing temperature is increased beyond 750°C, a significant blue-shift of the PL emission peak to 1.20eV is observed as noted from figure 1. This is probably due to the enhancement of In/Ga interdiffusion in the QDs at elevated temperatures surpassing the effects of vertical strain field propagation due to QD layer stacking and the indium atom gradient in the capping layer due to presence of quaternary alloy. Thus, with increased In/Ga interdiffusion the strain-relaxed state of the QDs is no longer preserved which causes the blue-shift of the PL emission peak. With further increase in annealing temperature to 850°C we observe PL emission peaks at shorter wavelength with reduced intensity, attributed to the dissolution of dots in the wetting layer.

Temperature dependent PL measurements are done to investigate the thermal quenching of the PL intensity in the annealed samples. The integrated PL intensity corresponding to the bound electron peak with increase in temperature is given by [20, 21]

$$I = \frac{I_0}{1 + C \exp(-\frac{E_A}{k_B T})}$$

(1)

where I_0 is the integrated PL intensity at 0 K, C is the ratio of the thermal escape rate to the radiative recombination rate, E_A is the activation energy and k_B is the Boltzmann constant. Higher activation energy (of the order of ~250 meV) obtained from calculations for the as-grown as well as 650°C annealed samples indicates stronger carrier confinement. For the samples annealed at 700°C and above, a steady decrease of the activation energy is observed, indicating weakening of the carrier confinement potential due to the increased In/Ga adatom diffusion in the QDs as explained. This may also be due to the generation of non radiative defect centers in the barriers as a result of increasing elastic stress due to annealing, which capture the thermally activated excitons and thereby reduce the activation energies of the carriers trapped in the QDs.

To investigate the dissolution of larger dots on high temperature annealing as ascribed from the PL study, we have taken the Raman spectrum for the annealed samples shown in figure 2. Though in Raman spectrum have some noise, for the 650°C annealed sample, an intense peak at 290 cm^{-1} is seen which corresponds to the GaAs longitudinal-optic (LO) phonon mode.

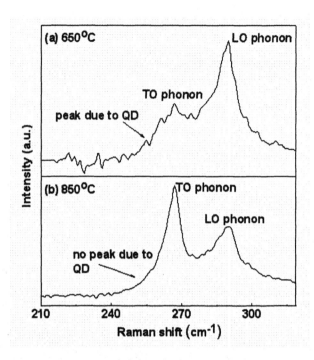

Figure 2. Raman spectroscopy of InAs/GaAs QDs annealed at (a) 650°C and (b) 850°C.

The relatively weak peak centered at 266 cm^{-1} is the transverse optic (TO) phonon of GaAs. The additional feature at a frequency lower than that of TO GaAs mode is attributed to the vibrational excitation of the InAs QDs [22, 23]. Disappearance of the QD phonon frequency is seen for the sample annealed at 850°C which is attributed to the complete dissolution of the larger dots. Similar report is also seen for AlSb QDs on GaAs substrate where Raman signal was observed from 4ML QD coverage but not from the 2ML QD coverage [24]. A strong signal is seen in the position of the GaAs TO phonon frequency for the sample annealed at 850°C. This probably occurred from the L⁻ plasmon mode of the n⁺ GaAs buffer layer [24], which may be due to the erosion of the InAs wetting layer on annealing.

To know the morphology and shape of QDs in as grown and annealed samples the TEM experiment has been carried out. Figure3. (a), (b) and (c) shows XTEM images of as grown and annealed samples at 750°C and 850°C respectively. It is clearly shown QDs are preserved even annealing up to 750°C temperature where complete dissolution of QDs occurs at 850°C temperature. The bump in figure (a) and (b) may be some artifact or small defect cause due to sample preparations or during growth.

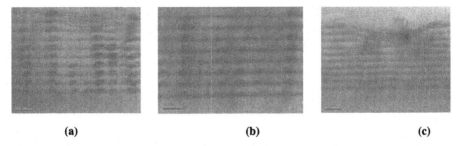

| (a) | (b) | (c) |

Figure 3. XTEM image of (a) as grown sample shows the formation of two different sizes of QDs. (b) annealed at 750°C indicates preservation of QDs at higher temperature and (c) annealed at 850°C shows dissolution of QDs in surrounding wetting layer.

Conclusion

In summary, we have investigated the post-growth annealing effects of the ten layer stacked InAs/GaAs QDs for asserting its thermal stability of peak emission wavelength suitable for lasers. The QD heterostructure exhibited almost no shift in the emission peak for annealing up to 700°C, contrary to the reports of pronounced blue-shifts on annealing at this temperature. Higher activation energy (of the order of ~250 meV) at 650°C annealing indicates its stronger carrier confinement potential at the elevated annealing temperature. Preservation of QDs also occurred up to 750°C as shown from XTEM images. These results thus demonstrate that this coupled multilayer stacked QD is a desirable alternative for better thermal stability of emission wavelength of optoelectronic devices.

Acknowledgement

The Authors acknowledge the financial support provided by the Department of Science and Technology, India. The authors also acknowledge the partial financial support provided by the Ministry of Communication and Information Technology, Government of India, through Centre of Excellence in Nanoelectronics and by the European Commission under contract SES6-CT-2003-502620 (FULLSPECTRUM).

References

[1] H. Y. Liu, B. Xu, Y. Q. Wei, D. Ding, J. J. Qian, Q. Han, J. B. Liang and Z. G. Wang, *Appl. Phys. Lett.* **79**, 2868 (2001).
[2] T. J. Badcock, R. J. Royce, D. J. Mowbray, M. S. Skolnick, H. Y. Liu, M. Hopkinson, K. M. Groom and Q. Jiang, *Appl. Phys. Lett.* **90**, 111102 (2007).
[3] S. Y. Wang, S. C. Chen, S. D. Lin, C. J. Lin and C. P. Lee, *Infrared Phys. & Technol.* **44**, 234 (2003).
[4] S. Chakrabarti, A. D. Stiff-Roberts, X. H. Su, P. Bhattacharya, G. Ariyawansa and A. G. U. Perera, *J. Phys. D: Appl. Phys.* **38**, 2135 (2005).

[5] M. Grundmann, N. N. Ledentsov, R. Heitz, L. Eckey, J. Christen, J. Böhrer, D. Bimberg, S. S. Ruvimov, P. Werner, U. Richter, J. Heydenreich, V. M. Ustinov, A. Y. Egorov, A. E. Zhukov, P. S. Kopev and Z. I. Alferov, *physica status solidi (b)* **188**, 585 (1995).

[6] P. G. Eliseev, H. Li, A. Stinz, G. T. Liu, T. C. Newell, K. J. Malloy and L. F. Lester, *Appl. Phys. Lett.* **77**, 262 (2000).

[7] N. Otsubo, M. Hatori, S. Ishida, T. Okumura, Y. Akiyama, H. Nakata, M. Ebe, A. Sugawara, and Y. Arakawa, *Jpn. J. Appl. Phys. Part2* **43**, L1124 (2004).

[8] A. Salhi, L. Fortunato, L. Martiradonna, M. T. Todaro, R. Cingolani, A. Passaseo and M. De Vittorio, *Semicond. Sci. Technol.* **22**, 396 (2007).

[9] H. Y. Liu, M. Hopkinson, K. Groom, R. A. Hogg and D. J. Mowbray, *Proc. SPIE* **6909**, 690903 (2008).

[10] P. M. Smowton, E. Herrmann, Y. Ning, H. D. Summers, P. Blood and M. Hopkinson, *Appl. Phys. Lett.* **78**, 2629 (2001).

[11] S. J. Xu, X. C. Wang, S. J. Chua, C. H. Wang, W. J. Fan, J. Jiang and X. G. Xie, *Appl. Phys. Lett.* **72**, 3335 (1998).

[12] H. S. Lee, J. Y. Lee, T. W. Kim and M. D. Kim, *J. Appl. Phys.* **94**, 6354 (2003).

[13] S. Adhikary, N. Halder, S. Chakrabarti, S. Majumdar, S. K. Ray, M. Herrera, M. Bonds and N. D. Browning, *J. Crystal Growth* **312**, 724 (2010).

[14] N. Halder, R. Rashmi, S. Chakrabarti, C.R. Stanley, M. Herrera, N.D. Browning, *Appl. Phys. A* **95**, 713 (2009).

[15] M. V. Maximov, A. F. Tsatsul'nikov, B. V. Volovik, D. S. Sizov, Yu. M. Shernyakov, I. N. Kaiander, A. E. Zhukov, A. R. Kovsh, S. S. Mikhrin, V. M. Ustinov, Zh. I. Alferov, R. Heitz, V. A. Shchukin, N. N. Ledentsov, D. Bimberg, Yu. G. Musikhin and W. Neumann, *Phys. Rev. B* **62**, 16671 (2000).

[16] S. Chakrabarti, N. Halder, S. Sengupta, J. Charthad, S. Ghosh and C. R. Stanley, *Nanotechnology* **19**, 505704 (2008).

[17] S. Chowdhury, S. Adhikary, N. Halder and S. Chakrabarti, *"A Novel Approach to increase emission wavelength of InAs/GaAs quantum dots by using a quaternary capping layer"* (to be published Opto-Electronics Review).

[18] M. Yu. Petrov, I. V. Ignatiev, S. V. Poltavtsev, A. Greilich, A. Bauschulte, D. R. Yakovlev and M. Bayer, *Phys. Rev. B* **78**, 045315 (2008).

[19] S. W. Ryu, I. Kim, B. D. Choe and W. G. Jeong, Appl. Phys. Lett. **67**, 1417 (1995).

[20] F. Y. Tsai and C. P. Lee, *J. Appl. Phys.* **84**, 2624 (1998).

[21] E. W. Williams and H. B. Bebb in *Semiconductor and Semimetals*, edited by R.K. Willardson, and A.C. Beer (vol. 8. New York: Academic Press; 1992) p. 321.

[22] J. Ibáñez, A. Patané, M. Henini, L. Eaves, S. Hernández, R. Cuscó, L Artús, Yu. G. Musikhin and P. N. Brounkov, *Appl. Phys. Lett.* **83**, 3069 (2003).

[23] L. Artús, R. Cusco, S. Hernández, A. Patanè, A. Polimeni, M. Henini and L. Eaves, *Appl. Phys. Lett.***77**, 3556 (2000).

[24] B. R. Bennett, B. V. Shanabrook and R. Magno, *Appl. Phys. Lett.* **68**, 958 (1996).

Mater. Res. Soc. Symp. Proc. Vol. 1272 © 2010 Materials Research Society   1272-LL09-11

Liquid State Polydimethylsiloxane (PDMS) Wrinkle Formation Process for Various
Applications.

Sang Hoon Lee[1], Jangbae Jeon[2], Sang Ho Lee[1], M. J. Kim[1]

[1]Department of Materials Science and Engineering, University of Texas at Dallas, TX75080,
USA
[2]Department of Electrical Engineering, University of Texas at Dallas, TX75080, USA

ABSTRACT

Here we report an epoch-making simple fabrication for wrinkle formation. The present
wrinkle formation process is a solution for controlling the area, shape and direction of wrinkle
area by forming wrinkles on the liquid state polydimethylsiloxane directly exposed to sputtered
metal particles in the low vacuum plasma chamber in various vacuum states and deposition
conditions. Also the process allows us to make extremely flexible metal thin film electrode with
approved adhesion. These bring us possibilities of actual electrical and biological applications.

INTRODUCTION

Applications of the present process could be applied to areas including electrode systems,
tissue engineering, cell behavior can be controlled by altering surface topology, cell culturing
substrate, including optical diffraction gratings and optical micro lenses, biosensors, cell weight
sensors, micro fluidic devices without mixers, and flexible and multi axial stretch able electrodes
[1 - 5]. Current wrinkle fabrication processes on solid state polydimethylsiloxane (PDMS) can
only make one directional wrinkle and cover a very tiny small area (micrometer square scale)[6].
The present method solves these problems by making wrinkles on the liquid state PDMS
using sputtered metal particles (ions) by plasma in the vacuum chamber. The present process
provides very good adhesion between the PDMS and the metal film layer, but is removable by
using etchant, has stretch ability, flexibility, and also can be made in any size of area in multi-
directional wrinkle formation. Increasing adhesion between the metal layer and the PDMS by a
heating process is not required. The process utilizes liquid state (or not cured) polymer and a
plasma sputtering system to make a broad vast expanse of wrinkled area and multi directional
wrinkle. All of the existing techniques are based on cured PDMS substrate. And this simple
process is only the liquid-state PDMS process.

THEORY

Bulk PDMS completely consists of a simple repeating structure ($-Si(CH_3)_2-O-$). But with
PDMS in the liquid state, this structure is not connected and repeated. Therefore the surface
energy of this liquid state is higher than bulk state PDMS, in other words, liquid state PDMS is

relatively less stable than solid state PDMS. Using this characteristic of liquid state PDMS, the process was developed to make a broad vast expanse of wrinkled area. Also it is possible to use low energy due to the ease of forming the wrinkled surface compared with high energy plasma or a high voltage ion beam. [1]

Relation between wave length (L) over hardened layer thickness (t) is related to the elastic moduli of film (E_f) and substrate(E_s) . Also amplitude (W) over thickness (t) relates applied strain (e) and critical strain (e_c), where critical strain is $e_c = 1/4(3E_s/E_f)^{2/3}$ [7]

$$L/t = \alpha\sqrt[3]{\overline{E}_f/\overline{E}_s} \qquad (1)$$

$$W/t = \sqrt{e/e_c - 1} \qquad (2)$$

Through this process, the strain in the hardened surface by sputtered metal particle and surface expansion difference can be estimated by measuring of surface length, L and (L-L_0)/L_0 can be considered as strain approximation [8]. In addition, the effect of pressure should be considered as variations of volume and surface area during liquid state wrinkle process .

EXPERIMENT

An exemplary embodiment of the present process can comprise the following materials. Gold (Au) for top film layer, Sylgard® 184 silicone elastomer base, Sylgard® 184 silicone elastomer curing agent, vacuum pump, desiccators, Denton Desk III TSC Sputter Coater. An exemplary embodiment of the present method may include the following steps. Mix a ratio of base to curing agent = 20:1, pour the mixed elastomer (PDMS) to 1 inch diameter plate or bigger plate, hold the mixed liquid state elastomer for about 70 to 80 min. in the low vacuum desiccators, put a plate into vacuum chamber (e.g., Denton Desk III TSC, but it is possible to use other vacuum chamber which can make a plasma metal sputter), keep the chamber in vacuum range of about 80 to 300 mTorr 5 using high purity (99.99%) argon gas, plasma sputtering gold (Au) deposition (time range from 90 to 135 seconds)and other metals instead of Gold (Au) can be used) in the argon gas filled plasma chamber. Film density setup is to be 16.38 g/cm3 (for Au plasma) or we can change to other metals for deposition. Plasma Impedance should be 23.80 Ohm (for Au plasma) or it is possible to change set values to other metals for deposition. After this process, vent the vacuum chamber directly, let the wrinkles form on surface of liquid state PDMS plate at room temperature for about 24 hours to cure it. And optionally, It is one of choices to remove gold film layer if it is desired to use just wrinkle on the PDMS by gold etchant. In this optional step, if it is desired to have just metal film coated onto the wrinkle surface PDMS substrate, the removal process is not necessary. But, if it is desired to have just a wrinkle surface PDMS substrate which is not containing gold metal film, then the gold film can be removed using gold etchant. Normal view of SEM and FIB assisted cross section SEM analysis of wrinkled PDMS were performed

108

DISCUSSION

Volume and surface of Liquid state PDMS in 80~120mTorr is increased in the vacuum chamber. Sputtered metal particle is penetrated in, diffused in and deposited on the jelly-like liquid state PDMS, At the same time the stiff thin layer is formed and this layer trigger the wrinkle formation. Relatively hardened surface thin film layer is formed on the area of the liquid state PDMS exposed to sputtered metal particles by plasma. Then a tendency for the hardened surface of PDMS thin film to expand in the every direction perpendicular to the direction of sputtered metal particle irradiation is occurred. The mismatch strain between the hardened surface thin film and jelly-like liquid state substrate gives rise to layer buckling and the formation of the wrinkle patterns in low vacuum chamber. During this process, the 1st order of wrinkles is made by buckling principle [9-12]. And then the surface of liquid state PDMS is continuously coated with metal thin film layers by sputtered metal particles. Since PDMS is not cured, as pressure is released to atmospheric pressure (the pressure is back to 760Torr), volume is reduced and surface with metal thin film layer begin to make higher orders of wrinkles because of reducing liquid state PDMS volume and shrinking surface simultaneously. After it cured, adhesion between gold film and wrinkled substrate is strong enough to endure adhesion test. It is proven as 'standard tape test' which is well known standard film adhesion test. In addition, we can get a broad vast expanse of multi axial wrinkled area during existing wrinkle formation process. Optionally, we can remove the Au layer using etchant to get bare wrinkled surface.

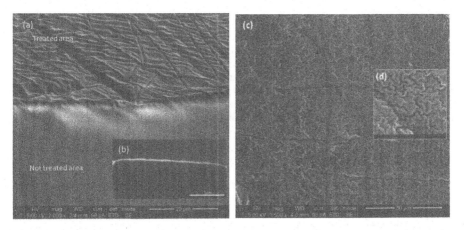

Figure 1. (a) Liquid state process treated area and not treated area by protecting mask (b) FIB assisted cross sectional SEM image. Top layer is Platinum (Pt). It shows surface morphology. (c) Wrinkle area physically peeled off gold (Au) thin layer and (d) Enlarged wrinkle image.

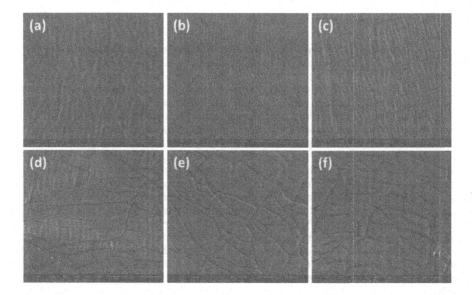

Figure 2. Vacuum pressure dependence of Wrinkle formation due to expansion and shrink of surface (a) 80~120mTorr (b) 140mTorr (c) 160mTorr (d) 180mTorr (e) 200mTorr (f) 300mTorr

At 120mTorr plasma range, stiff film layers have an improved adhesion property on polydimethylsiloxane substrate. 8 mm × 30 mm Au films are made on the wrinkled surface of elastomer before curing the polydimethylsiloxane. Figure 2 presents a set of controllable wrinkle formations. As formation pressure is increased, stiff surface tends to make different type of higher order wrinkles on the surface. It can be explained that unstable energy state on surface such as induced strain by lattice mismatch. Different pressure conditions lead to different wrinkle patterns and it needs more specified analysis to understand mechanism.

Applications and Simulation

Most recent applications of these wrinkle patterns are stretchable electrode and Lab on a chip. Here I report one of prospective data which is random wrinkle pattern in micro fluidic channel. Comparison of two-dimensional turbulence simulation with periodic shape of pattern, regular wrinkle pattern and irregular wrinkle pattern is performed. And irregular wrinkle pattern tends to cause more turbulence than all. Arrows on Figure 3 is the direction of flux. Inlet constant velocity profile in the x direction is set to 0.0001 m/s and velocity in the y direction is zero. And it is applied a zero-pressure condition to the outlet boundary. Also at all other boundaries, logarithmic wall functions are applied as boundary conditions.

110

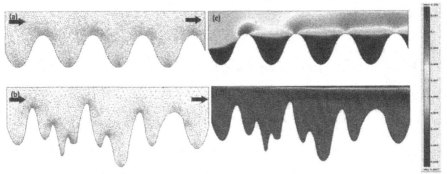

Figure 3. 2-D Turbulence FEM simulation of wrinkled channel in micro fluidic device. (a)mesh design of regualer wrinkel (b) mesh design of irregular mesh design (c) Turbulence at regular wrinkle induces device (d) Turbulence at irregular wrinkle induces device

CONCLUSIONS

This wrinkle process suggests many feasible solutions for the 'size matter' in electrical and biological applications such as flexible, stretchable electrodes, sensor, cell culturing substrates and nano/micro scale fluidic mixing channels. If one desires to have bigger size of wrinkle surface PDMS substrate, he/she can make a large vacuum chamber and put a large size plate containing the liquid state PDMS with a larger vacuum pump to make same circumstance in the present process.

The rheological properties of liquid state PDMS enables to develop novel and simple process for making a broad vast expanse of wrinkled area which has never been made before. In addition, comparing with high energy plasma treatment methods, ultraviolet/ozone (UVO) or high voltage ion beam process, jelly-like liquid state PDMS enables the relatively lower energy consumption due to its higher surface energy to form the wrinkled surface.

This simplified process, comparing with existing process methods, enables to make controllable huge wrinkle formation area. This process can make even centimeter square scale area now and bigger wrinkled area on the targeted elastomer substrate can be made by research which is on the way.

ACKNOWLEDGMENTS

This project is supported through the program of FUSION by The Texas Emerging Technology Fund (TETF) of the Texas State Government, United States of America.

REFERENCES

1. M.-W. Moon, S.H. Lee, J.-Y. Sun, K.H. Oh, A. Vaziri, J.W. Hutchinson, *Proc. Natl Acad. Sci. USA* 104 (2007) 1130
2. A. Khademhosseini, R. Langer, J. Borenstein, J. P. Vacanti, *Proc. Natl Acad. Sci. USA* 103 (2006) 2480
3. X. Jiang, S. Takayama, X. Qian, E. Ostuni, H. Wu, N. Bowden, P. LeDuc, D.E. Ingber, G.M. Whitesides, *Langmuir* 18 (2002) 3273.
4. A.K. Geim, S.V. Dubonos, I.V. Grigorieva, K.S. Novoselov, A.A. Zhukov, S.Y. Shapoval, *Nature Mater.* 2 (2003) 461
5. Ju-Yeoul Baeka,c, Jin-Hee Ana, Jong-Min Choi, Kwang-Suk Park, Sang-Hoon Lee *Sen. Actu. A* 143 (2008) 423–429
6. Alexandra Schweikart, Andreas Fery, *Microchim Acta* 165 (2009) 249-263
7. S. Chung, J. H. Lee, M.-W. Moon, J. Han, Roger D. Kamm, *Adv. Mater.* (2008) 20, 3011–3016
8. M.-W. Moon, S.H. Lee, J.-Y. Sun, K.H. Oh, A. Vaziri, J.W. Hutchinson, *Scrip. Mater* 57 (2007) 1130
9. Bowden N, Brittain S, Evans AG, Hutchinson JW, Whitesides GM (1998) *Nature* 393:146–149
10. K. Efimenko, M. Rackaitis, E. Manias, A. Vaziri, L. Mahadevan, J. Genzer, *Nat. Mater.* 4 (2005) 293.
11. Kim YR, Chen P, Aziz MJ, Branton D, Vlassak JJ(2006) *J Appl Phys* 100:104322.
12. E. Cerda, L. Mahadevan, *Phys. Rev. Lett.* (2003) 90, 074302

Mater. Res. Soc. Symp. Proc. Vol. 1272 © 2010 Materials Research Society   1272-LL11-08

Controlled Stepwise Growth of Siloxane Chains Using Bivalent Building Units With Different Functionalities

Nils Salingue, Dominic Lingenfelser, Pavel Prunici, and Peter Hess

Institute of Physical Chemistry, University of Heidelberg

Im Neuenheimer Feld 253, D-69120 Heidelberg, Germany

ABSTRACT

Organic/inorganic hybrids of silicon and their subsequent chemical modification are of interest for tailoring and structuring surfaces on the nanoscale. The formation of monolayers on hydroxylated silicon surfaces was employed to synthesize molecular dimethylsiloxane chains by wet-chemical condensation reactions, using dimethylmonochlorosilane as the precursor. The SiH group of the resulting dimethylsilyl termination could be selectively oxidized to the SiOH group, which opened the possibility of bonding another species. By repeating the condensation and oxidation cycle the stepwise growth of one-dimensional dimethylsiloxane chains was achieved. The ongoing chain growth was characterized by attenuated total reflection (ATR) Fourier transform infrared (FTIR) spectroscopy, x-ray photoelectron spectroscopy (XPS), spectroscopic ellipsometry (SE), and determination of the surface energy by contact-angle experiments.

INTRODUCTION

Poly(dimethylsiloxane) (PDMS) is a polymeric material with growing applications due to its excellent properties such as biocompatibility, nontoxicity, optical transparency, low surface energy, hydrophobicity, robust backbone, high flexibility, and excellent thermal properties [1]. Therefore, besides many established practical applications, PDMS is also of increasing interest in current research, for example, in bio-microelectromechanical system (bio-MEMS) devices for microfluidics to control the nanoscale properties of the surface (labs-on-a-chip) [2]. PDMS is also of interest in microcontact printing (μCP), offering functionalization and thus extension to the use of polar molecules as inks by surface treatment and low cost surface patterning [3]. To achieve these goals it is necessary to carefully control and tailor the physical and chemical surface properties of thin layers of PDMS. A more detailed discussion of the field of surface-functionalized silicone elastomer networks can be found in [4]. It is important to note that several of the reported properties are superior to those of self-assembled monolayers (SAMs) of long-chain alkyl silanes,which may also be used in microfluidics, cell growth, and soft lithography.

Despite the needs to understand and control the surface chemistry on the molecular level, results on the controlled formation of siloxane monolayers and the elucidation of their properties have been reported only in very few publications. This includes the chemisorption of monolayers of 1,3,5,7-tetramethylcyclotetrasiloxane (TMCTS) on flat substrates of TiO_2 [5]. In this work sequential cycles of TMCTS chemisorption and the photoinduced oxidation to uniform SiO_x

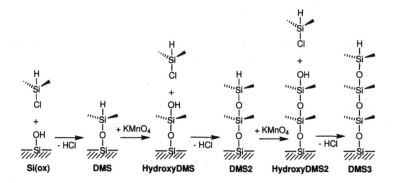

Figure 1. Schematic representation of the first five steps of the bottom-up synthesis of one-dimensional chains of dimethylsiloxane on oxidized silicon.

layers were studied. However, the nonselective simultaneous UV photo-oxidation of the SiH and SiCH$_3$ groups of TMCTS generates different reaction centers in each siloxane unit. This makes it extremely difficult to realize well-ordered chain growth with this kind of molecular building blocks.

Oligo(dimethylsiloxane) (ODMS)-covered surfaces of mesoporous precipitated silica have been prepared by reactions of bivalent Cl(Si(CH$_3$)$_2$O)$_n$Si(CH$_3$)$_2$Cl, n = 0,1,2,3,4 precursors with wet silicas [6]. In this particular work the bonding density of siloxane species at the surface and the dependence of covalent bonding on the amount of preadsorbed water was investigated. The disadvantage of these bifunctional siloxanes, with two reactive SiCl groups, is that a controlled stepwise growth of linear siloxane chains is not possible, due to their tendency to form cyclic structures and the fact that already the smallest dimethylsiloxane (DMS) unit (n = 0) may be bonded to the surface by both ends [6]. The reaction with dry silicas yielded essentially a single layer of poly(dimethylsiloxane) loops covalently attached to the surface, while the coverage increased with water content. Note that the actual distribution of chain length varied substantially with preparation conditions. Thus, the occurrence of looped and extended configurations renders controlled 1 D chain growth impossible.

The goal of the work presented here was to develop a method that allows for the first time a controlled stepwise growth of siloxane chains on hydroxylated silicon, as illustrated in Fig. 1. For this purpose dimethylmonochlorosilane ((CH$_3$)$_2$HSiCl) was employed that contains besides a very reactive SiCl group a much less reactive SiH functionality. Since the latent reactivity of the SiH bonds must be triggered by a secondary selective activation step, e.g., oxidation of SiH to SiOH, the chemical reactions of the two functionalities can be separated and well-defined chain growth could be achieved on silicon in a step-by-step manner.

Analysis of the various growth steps studied by attenuated total reflection (ATR) Fourier transform infrared (FTIR) spectroscopy, x-ray photoelectron spectroscopy (XPS), determination of the surface energy by contact-angle experiments, and spectroscopic ellipsometry (SE: NIR-UV) clearly indicates that in this case the structural inhomogeneity and chemical variety , usually encountered in a bottom-up growth process with bifunctional and trifunctional precursors, can be avoided.

114

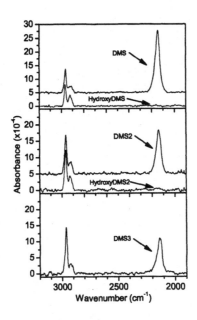

Figure 2. ATR-FTIR spectra of DMS, hydroxyDMS, DMS2, hydroxyDMS2, and DMS3 with the symmetric and antisymmetric CH stretching vibrations and the SiH stretching mode.

EXPERIMENT

The dimethylsiloxane (DMS)-terminated surface was prepared by covering the surface with a thin liquid film of pure 1,1,3,3-tetramethyldisilazane directly on the oxidized silicon surface. After 45 s the reagent was evaporated at room temperature and the crystal was rinsed with heptane and acetone. Alternatively, this termination can be prepared by using a solution of dimethylmonochlorosilane in heptane (1:50, v/v) at room temperature for 10 min.

For the selective oxidation of SiH to SiOH in DMS with potassium permanganate, the sample was dipped for 30 min in a 0.01 M aqueous KMnO$_4$ solution. Finally the oxidized surface with hydroxyDMS was rinsed with water and cleaned in acetone by ultrasonic waves. This should be compared with the conventional oxygen plasma treatment [7] and photoinduced UV/ozone oxidation [8]. These procedures modify PDMS from a hydrophobic methyl-terminated surface to a hydrophilic hydroxyl-terminated surface and are also suitable for patterning [9].

For the synthesis of layers of tetramethyldisiloxane (DMS2) and hexamethyltrisiloxane (DMS3) the cleaned hydroxyDMS or hydroxyDMS2 surface was dipped into a solution of dimethylmonochlorosilane in heptane (1:50, v/v) at room temperature. After a reaction time of 10 min the sample was removed and rinsed with heptane and acetone. Alternatively, the DMS2 (DMS3) surface can be synthesized, as described above for DMS, by applying 1,1,3,3-tetramethyldisilazane directly onto the hydroxyDMS (hydroxyDMS2) terminated surface.

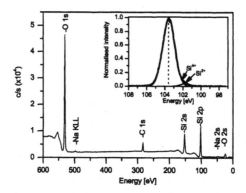

Figure 3. XPS survey spectrum and resolved Si 2p spectrum (inset) of the DMS2 termination.

RESULTS AND DISCUSSION

The stepwise growth of the 1D siloxane chains is illustrated schematically in Fig. 1, up to the third repeat unit. The following detection methods used to analyze the first processing steps indicate that with bifunctional building blocks, containing a reactive (SiCl) and a less reactive functional group (SiH) well-ordered chain growth can be realized [10,11].

The reaction processes were characterized by ATR-FTIR spectroscopy using a suitable silicon crystal. As can be seen in Fig. 2, the intensity of the CH signals, caused by the symmetric and antisymmetric stretching vibrations increased during chain growth owing to the larger number of methyl groups. The peak of the SiH stretching vibration disappeared completely during oxidation and reappeared slightly shifted, due to the changed chain length, after the next condensation reaction. Obviously, as a consequence of a reduced yield in the second and third growth steps, the intensity of the SiH peak did not reach the same intensity as in the first step. This loss effect can also be seen in the CH signals. Nevertheless, the spectroscopic results suggest unit-by-unit extension of 1D siloxane chains by directed assembly, based on a monochlsilane as precursor.

To further circumstantiate chain growth, XPS measurements on samples with different siloxane chain lengths were performed. The survey spectrum shows, besides the strong O 1s, a small C 1s, and the two Si 1s and Si 2p peaks, no unexpected peaks (see Fig. 3). The enlarged Si 2p peak in the inset exhibits besides the dominating contribution of bulk Si^{4+} at 103.5 eV, a small shoulder belonging to Si^{2+} of the siloxane chains. To substantiate chain growth the small Si^{2+} peak was integrated for samples prepared with increasing number of repeat units. The observed linear increase of the Si^{2+} signal with the number of building blocks clearly suggests step-by-step chain growth by the applied sequence of condensation and oxidation cycles.

To support the selective oxidation of the SiH bonds and stepwise chain growth, contact-angle measurements with five different liquids were performed to determine the surface energy. The largest change occurred in the polar part of the surface energy, probably due to the polar SiOH end group, which is able to form H bonds with some of the test liquids. When the SiH group of DMS was oxidized to hydroxyDMS, the polar part increased from 1.7 to 9.5 mN/m, whereas the dispersive part changed only from 25.1 to 26.1 mN/m (see Fig. 4). This is consistent with the

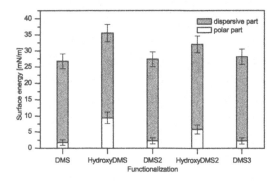

Figure 4. Variation of the dispersive and polar parts of the surface energy for the selective oxidation of DMS and DMS2 and for DMS3, as extracted from contac-angle measurements.

efficient oxidation of the nonpolar SiH to the polar SiOH group. After the conversion of hydroxyDMS to DMS2 the polar part decreased, whereas it increased again from 2.3 to 5.9 mN/m after oxidation of DMS2 to hydroxyDMS2. The measured water contact angles were 97° for DMS, 70° for hydroxyDMS, 92° for DMS2, 82° for hydroxyDMS2, and 94° for DMS3.

Real-time spectroscopic ellipsometry (SE) allows a highly sensitive and quite accurate determination of the thickness of monolayers if this layer can be removed in situ, e.g., by complete oxidation of a hydrocarbon termination to volatile products by laser irradiation [12]. The sensitivity of this method is sufficient to measure not only the thickness of C18 SAMs (2.3 nm) very accurately but also the effective thickness change during oxidation of the DMS monolayer. In the latter oxidation process only the two methyl groups and the hydrogen will disappear, while the SiO chain may be further oxidized but remains on the surface. Extensive oxidation and rearrangement of the interface may even lead to a silicon-oxide surface layer. For the DMS monolayer, spectroscopic ellipsometry yielded a thickness change of 0.24 nm, which is in the same range as the van der Waals radius of methyl groups. Since the SiO backbone of the longer siloxane chains cannot be removed by laser-induced oxidation, their real chain lengths could not be determined. In fact, the oxidative removal of the methyl groups and the oxidative rearrangement of the SiO backbone resulted in a measured thickness change of only 0.36 nm for DMS2 and 0.44 nm for DMS3, while the real length of DMS3 is about 1 nm.

The thicknesses of the siloxane layers considered here are much smaller than those of the ultrathin poly(siloxane) films of 3-5 nm described previously [13]. Figure 5 provides a comparison of the real-time ellipsometric thickness measurement of a C18 monolayer, prepared with octadecyltrichlorosilane (self assembly), a C20 monolayer, prepared with octadecyl-dimethylmonochlorosilane (directed assembly), and a DMS monolayer (directed assembly).

CONCLUSIONS

In the work presented here, the diversity of reaction schemes leading to spatial organization at a surface is substantially extended, providing access to desired bivalent functionalities on the monolayer scale for a class of precursor molecules containing SiH and SiCl bonds. The applied

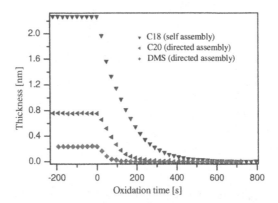

Figure 5. Ellipsometric thickness measurements of the hydrocarbon part of a C18 monolayer (self- assembly), C20 monolayer (directed assembly), and DMS (directed assembly).

concept of using functionalities with different reactivity, one that reacts immediately and a latent one that first must be activated, enormously increases the control of surface chemistry at the monolayer level. The resulting arrangement of small molecular units by directed assembly seems to be an effective way of realizing solid-supported architectures based on the well-defined organization of small building blocks. Siloxanes already serve as robust coupling agents between organic and inorganic materials and have found application to protect the surface and hinder detrimental reactions. The activated two-step-growth mechanism opens new avenues not only in siloxane surface chemistry. Therefore, this approach contributes to the ongoing interest in finding coupling agents that allow a selective growth of nanostructures and a change in the chemistry of the end groups on demand.

REFERENCES

1. N. Auner and J. Weis (Eds.), *Organosilicon Chemistry IV: From Molecules to Materials*, Wiley-VCH, Weinheim, 2000.
2. I. Wong and C.-M. Ho, *Microfluidics Nanofluidics* **7**, 291 (2009).
3. A. Perl, D. N. Reinhoudt, and J. Huskens, *Adv. Mater.* **21**, 2257 (2009).
4. J. A. Crowe, K. Efimenko, and J. Genzer, *ACS Symp. Ser.* **964**, 222 (2007).
5. H. Tada, *Langmuir* **12**, 966 (1996).
6. A. Y. Fadeev and Y. V. Kazakevich, *Langmuir* **18**, 2665 (2002).
7. H. Hillborg, J. F. Ankner, U. W. Gedde, G. D. Smith, H. K. Yasuda, and K. Wikstrom, *Polymer* **41**, 6851 (2000).
8. F. Egitto and L. Matienzo, *J. Mater. Sci.* **41**, 6362 (2006).
9. H. Ye, Z. Gu and D. H. Gracias, *Langmuir* **22**, 1863 (2006).
10. N. Salingue, D. Lingenfelser, P. Prunici, and P. Hess, unpublished.
11. N. Salingue, D. Lingenfelser, P. Prunici, and P. Hess, *SPIE Proc.* **7364**, 73640F-1 (2009).
12. P. Prunici and P. Hess, *J. Appl. Phys.* **103**, 024312 (2008).
13. M. K. N. Hirayama, W. R. Caseri, and U. W. Suter, *J. Colloid. Interface Sci.* **216**, 250 (1999)

Mater. Res. Soc. Symp. Proc. Vol. 1272 © 2010 Materials Research Society   1272-LL11-12

Mediating the Hydrogen Bonding Strength Modulates the Phase Behavior of Block Copolymer/Homopolymer Blends

Shih-Chien Chen[a], Shiao-Wei Kuo[b], U-Ser Jeng[c], Chun-Jen Su[c], and Feng-Chih Chang[a*]

[a]Institute of Applied Chemistry, National Chiao Tung University , Hsin Chu, 300 Taiwan

[b]Department of Materials and Optoelectronic Science, Center for Nanoscience and Nanotechnology, National Sun Yat-Sen University, Kaohsiung, 804, Taiwan

[c]National Synchrotron Radiation Research Center, Hsin Chu Science Park, Taiwan

ABSTRACT

We have investigated the phase behavior of two A-b-B/C blend systems, poly(4-vinylphenol-b-styrene)/poly(4-vinylpyridine) (PVPh-b-PS/P4VP) and poly(4-vinylphenol-b-styrene)/ poly(methyl methacrylate) (PVPh-b-PS/PMMA), featuring two types of hydrogen-bond strengths, and compared it with that of the A-b-B/A blend system poly(4-vinylphenol-b-styrene)/poly(4-vinylphenol) (PVPh-b-PS/PVPh). The PVPh-b-PS/P4VP blend system exhibited a series of micro-phase separation transitions, due to a decrease in the free energy, upon with increasing the P4VP content. In contrast, both the PVPh-b-PS/PVPh and PVPh-b-PS/PMMA blend systems maintained their lamellar structures upon increasing the PVPh or PMMA homopolymer content but tended to lose their long-range order. The introduction of an attractive interaction unit could effectively promote the miscibility of polymer blends. In an A-b-B/C blend system featuring hydrogen bonding interactions between the homopolymer and copolymer, the C homopolymer tends to form a wet brush-like structure and brings the morphological transitions (e.g., for the PVPh-b-PS/P4VP blend system) when the inter-association equilibrium constant (K_A) is greater than the self-association equilibrium constant (K_B). In contrast, the C homopolymer cannot bring the morphological transitions when K_A is less than K_B (e.g., for the PVPh-b-PS/PMMA blend system).

INTRODUCTION

Self-assembly is a key aspect in the design of many new functional supramolecular materials. Block copolymers are especially good candidates for producing self-assembled materials and nanometer-scale devices having a wide range of structural applications-for example, in advanced material formation, pollution control, and drug delivery.[1,2] Blending diblock copolymers (A-b-B) with homopolymers have attracted great interest in polymer science during recent decades because of their unusual phase behavior.[3]

In this study, we focused our attention on investigating the first case because only a few morphological studies have dealt with the influence of attractive interactions on micro-phase–separated structures. Indeed, the phase behavior of A-b-B/C blend systems with different attractive interaction strengths has not been established systematically. Our objective in this study was to synthesize a suitable A-*b*-B diblock copolymer (i.e., PS-*b*-PVPh) and investigate its behavior when blended with two different hydrogen bond–acceptor polymers that possess different hydrogen bonding strengths for their interactions with PVPh. The hydrogen-bonded polymer blends PVPh/P4VP and PVPh/PMMA are well-established as miscible blends.[4-6] Based on the Painter–Coleman association model (PCAM), the inter-association equilibrium constants for PVPh/P4VP (K_A = 598) and PVPh/PMMA (K_A = 37.4) and the self-association equilibrium constant for PVPh (K_B = 66.8) have been reported.[6] As a result, we established a new parameter K_A/K_B (i.e., K_A/K_B > 1 for PVPh/P4VP, K_A/K_B < 1 for PVPh/PMMA) to judge the relative hydrogen bonding strength and chain behavior of homopolymers in diblock copolymer/homopolymer blend systems. For comparison with the A-*b*-B/C blend, we also prepare a series of A-*b*-B/A (PVPh-*b*-PS/PVPh) blends. In this study, we used transmission electron microscopy to characterize the phase separation behaviors, and spatial distributions of the added homopolymers in A-*b*-B/C blend systems, featuring different specific interactions, as a function of the blend volume ratio.

EXPERIMENT

Block Copolymer and Homopolymer Syntheses

The PVPh-*b*-PS diblock copolymer and the PVPh, P4VP, and PMMA homopolymers were synthesized through sequential anionic polymerization as described in the previous study; their molecular weights are summarized in Table 1.[4, 7]

Table 1. Characterization of PVPh-*b*-PS, PVPh, PMMA, and P4VP Prepared by Anionic Polymerization and PVP

Polymer	Mn	Mw/Mn	Volume Fraction of PVPh (%)	Volume Fraction of PS (%)
PVPh$_{63}$-b-PS$_{109}$ (HS)	18900	1.07	37.6	62.4
PVPh$_{42}$ (H)	5040	1.05		
PMMA$_{53}$ (M)	5300	1.18		
P4VP$_{52}$ (V)	5460	1.12		

Blend Preparation

Blends of various PVPh-*b*-PS/P4VP (denoted HS/V), PVPh-*b*-PS/PMMA (denoted HS/M), and PVPh-*b*-PS/PVPh (denoted HS/H) compositions were prepared through solution casting. A DMF solution containing 5 wt% polymer mixture was stirred for 6–8 h and then cast on a Teflon dish. The solution was left to evaporate slowly at 100 °C for 7 days and then the sample was annealed under vacuum at 120 °C for 7 days. We have defined inter and intra chain association as shown in Scheme 1.

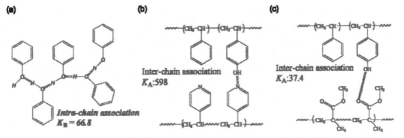

(a)   (b)   (c)

Inter-chain association K_A:598

Inter-chain association K_A:37.4

Intra-chain association $K_B = 66.8$

Scheme 1. The(a) intra-chain association of PVPh segments and inter-chain associations of (b) PVPh-*b*-PS/P4VP and (c) PVPh-*b*-PS/PMMA blend system were defined.

Characterization

The samples were then stained with RuO₄ (to stain PS) or I₂ (to stain P4VP) and viewed using a Hitachi H-7500 transmission electron microscope operated with an accelerating voltage of 100 kV. Ultrathin sections of the sample were prepared using a Leica Ultracut UCT microtome equipped with a diamond knife. Slices of ca. 70 nm thickness were cut at room temperature.

DISCUSSION

Figure 1 displays TEM images of the morphologies of the pure PVPh-*b*-PS diblock copolymer and the HS/V blend system with various compositions, after staining with I₂ for 24 h. The HS/V system undergoes a series of phase transitions upon increasing the P4VP content (i.e., lamellar → bicontinuous structure → hexagonal packed cylinders → BCC spheres).

Figure 1. TEM images of the morphologies of the solution-cast films of (a) pure HS and of the blends (b) HS/V = 90/10, (c) HS/V = 78/22, (d) HS/V = 71/29, and (e) HS/V = 29/71.

Figure 2 presents TEM micrographs obtained from the HS/M blend system with different compositions. The darker regions correspond to the PS-rich domains (stained with RuO₄) and the bright regions correspond to the mixed domains of PMMA and PVPh. The lamellar structure of the micro-phase separation of this system was maintained even when the PMMA content was large, indicating that the morphology of this system was unaffected by the solubilized PMMA.

121

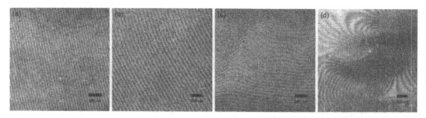

Figure 2. TEM images of the morphologies of the solution-cast films of the blends (a) HS/M = 90/10, (b) HS/M = 79/21, (c) HS/M = 62/38, and (d) HS/M = 30/70

The TEM images in Figure 3 reveal that all of the HS/H systems (various PVPh homopolymer contents) exhibit lamellar microstructures. The preservation of the lamellar morphology in this HS/H system indicates that (i) the penetration of the solubilized PVPh into the brush formed by the PVPh block chains was insignificant and (ii) these PVPh block chains formed the dry brushes. Consequently, the lamellar morphology of the PVPh-*b*-PS block copolymer was unaffected by the amount of added PVPh homopolymer, due to the dry brush behavior of the PVPh homopolymer.

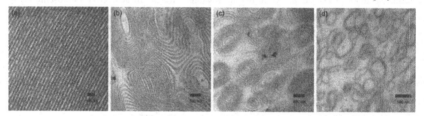

Figure 3 TEM images of the morphologies of the solution-cast films of the blends (a) HS/H = 90/10, (b) HS/H = 80/20, (c) HS/H = 63/37, and (d) HS/H = 31/69

In summary, we have established a novel parameter, K_A/K_B, for predicting the phase behavior of A-*b*-B/C blend systems featuring hydrogen bonding interactions between the C homopolymer and the A block of copolymer, where the ratio of the molar weight of the C homopolymer ($M_{C,homo}$) to the A block of the copolymer ($M_{A,block}$) is close to unity; when K_A/K_B is greater than unity (i.e., in the HS/V blend system), C has solubility limit within the A domains; in contrast, when K_A/K_B is less than unity (i.e., in the HS/M blend system), C is less uniformly solubilized within the A domains. As a result, we conclude that C brings the morphological transitions when K_A/K_B is greater than unity, whereas it cannot bring the morphological transitions when K_A/K_B is less than unity.

CONCLUSIONS

We have used TEM to investigate the phase behavior of HS/V and HS/M blend systems featuring various hydrogen bonding strengths between the homopolymers and diblock copolymers and compared the behavior with that of the HS/H blend system. TEM analyses both indicated that the HS/V blend system underwent a series of phase transitions upon increasing the P4VP content, due to decreases in the free energy in the system; in contrast, the HS/H and HS/M blend systems both maintained their lamellar structures upon increasing the PVPh and PMMA homopolymer contents, but lost their long-range order. For an A-b-B/C blend system featuring hydrogen bonding interactions between the homopolymers and copolymer, when the ratio of the molar weight of the C homopolymer ($M_{C,homo}$) to the A block of the copolymer ($M_{B,block}$) is close to unity and when K_A/K_B is greater than unity, the C homopolymer tends to form wet brush–like structures and brings the morphological transitions, such as in the HS/V blend system. In contrast, the C homopolymer cannot bring the morphological transitions, such as in HS/M blend system, when K_A/K_B is less than unity.

REFERENCES

1. Muthukumar, M.; Ober, C. K.; Thomas, E. L. *Science* **1997**, *277*, 1225.
2. Stupp, S. I.; Braun, P. V. *Science* **1997**, *277*, 1242.
3. Hashimoto, T.; Tanaka, H.; Hasegawa, H. *Macromolecules* **1990**, *23*, 4378. Tanaka, T.; Hasegawa, H.; Hashimoto, T. *Macromolecules* **1991**, *24*, 240.
4. Lin, C.-L.; Chen, W.-C.; Liao, C. S.; Su, Y. C.; Huang, C. F.; Kuo, S. W.; Chang, F. C. *Macromolecules* **2005**, *38*, 6435.
5. Chen, S. C.; Kuo, S. W.; Jeng, U S.; Su, C. J.; Feng-Chih Chang, F. C. *Macromolecules* **2010**, *43*, 1083.
6. Coleman, M. M.; Graf, J. F.; Painter, P. C. *Specific Interactions and the Miscibility of Polymer Blends*, Technomic Publishing: Lancaster, PA, **1991**.
7. Chen, S. C.; Kuo, S. W.; Liao, C. S.; Chang, F. C. *Macromolecules* **2008**, *41*, 8865

Materials Exploiting Peptide and Protein Self Assembly— Toward Design Rules

Mater. Res. Soc. Symp. Proc. Vol 1272 © 2010 Materials Research Society   1272-NN05-04

Enzyme Triggered Gelation of Arginine Containing Ionic-Peptides

Jean-Baptiste Guilbaud[1], Aline F. Miller[2] and Alberto Saiani[1,*]

[1]School of Materials, University of Manchester, Grosvenor Street, Manchester M1 7HS, UK.
[2]School of Chemical Engineering and Analytical Sciences and Manchester Interdisciplinary Biocentre, University of Manchester, 131 Princess Street, Manchester M1 7DN, UK.
* corresponding author: E-mail: a.saiani@manchester.ac.uk.

ABSTRACT

We have investigated the possibility of using the protease enzyme thermolysin to catalyse the synthesis and gelation of ionic-complementary peptides from non-gelling peptide precursors. In the described system, thermolysin was added at a fixed concentration (0.3 mg mL^{-1}) to solutions (25 - 100 mg mL^{-1}) of a short tetra-peptide FEFR. Initially, the protease partially hydrolysed the tetrapeptide into di-peptides in all samples. Subsequently, longer peptide sequences were found to form through reverse-hydrolysis and their stability was found to be dependent on their self-assembling properties. The sequences that self-assembled into anti-parallel β-sheet rich fibres became the stable products for the reverse hydrolysis reaction, while the others formed were unstable and disappeared with increasing incubation time. Ultimately, the main product of the system was octa-peptide, which suggests that it represents the thermodynamically favoured product of this dynamic library.

INTRODUCTION

Molecular self-assembly has emerged as a powerful tool for the fabrication of molecular materials with a wide variety of properties. In recent years, considerable advances have been made in using simple, synthetic oligo-peptides as building blocks due to the propensity of such building block to self assemble into ordered supramolecular structure for the production of novel biomaterials. Of particular interest, being able to trigger the self assembly of these small molecules by an external stimulus such as enzyme, light, pH or ionic strength is attracting increasing attention as a route for the reversible fabrication of soft solid biomaterials due to their potential applications in drug delivery, bio-sensing or regenerative medicine. We have investigated the possibility of using the protease enzyme thermolysin to catalyse the synthesis of ionic complementary oligo-peptides *via* reverse hydrolysis, with subsequent trigger of their gelation. Such ionic oligo-peptides are known to readily self-assemble into β-sheet affording rich fibrillar hydrogels [1-4] and are gaining increasing popularity in the literature due to their potential biomaterials applications and for aiding the understanding of the general paradigms that govern molecular self-assembly. We successfully applied this approach starting from the tetra-peptide FEFK as the related octa-peptides FEFKFEFK and FEFEFKFK readily self-assemble into β-sheet rich fibers to form self-supported hydrogels at low concentrations (10 and 15 mg ml^{-1} respectively).[5] Herein, this method was adapted using the tetra-peptide FEFR to fabricate stable gels as the poorly soluble related oligo-peptides based on FE and FR (FEFEFRFR and FEFRFEFR) form unstable gels at low concentration (ca. 5 mg mL^{-1}).

EXPERIMENT

FEFR (F is phenylalanine, E is glutamic acid and R is arginine, Figure 1) solutions with initial concentration, C_0, ranging from 25 to 100 mg mL^{-1}, were prepared in distilled water and their pH adjusted to 7. Thermolysin from *Bacillus thermoproteolyticus rokko* (Sigma Aldrich) -

known to hydrolyze peptide bonds on the amino side of hydrophobic residues [6], in our case F, was added at a fixed concentration (0.3 mg mL^{-1}) to the tetra-peptide solutions. The resulting samples were used immediately for analysis. The enzyme triggered gelation of FEFR was investigated initially using oscillatory rheometry (strain-controlled Bohlin C-CVO rheometer) as a function of time to confirm that a gel does form and to identify the critical gelation time. Subsequently matrix assisted laser desorption ionisation time of flight mass spectroscopy (MALDI-TOF MS, Axima CFR mass spectrometer, Shimadzu Biotech) in conjunction with reverse phase high performance liquid chromatography (RP-HPLC, Ultimate 3000 system with variable wavelength UV detector, Dionex) were used to both qualitatively and quantitatively characterize the peptide sequences synthesized at different time points. The development of the secondary structure of the peptide was monitored using time-resolved Fourier transform infra-red spectroscopy (FTIR, Thermo Nicolet 5700 spectrometer).

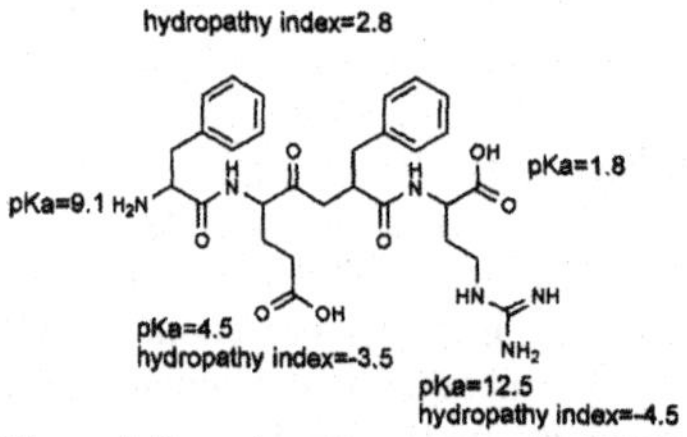

hydropathy index=2.8

pKa=9.1 H$_2$N

OH pKa=1.8

pKa=4.5
hydropathy index=-3.5

HN   NH

NH$_2$

pKa=12.5
hydropathy index=-4.5

Figure 1. Formula of the tetra-peptide FEFR showing the pKa and hydropathy index of the amino acid residues.

RESULTS AND DISCUSSION

Kinetics of gelation and rheological properties

No gel was found to form using the "tilting test-tube method" neither an increase in sample viscosity was observed for pure tetra-peptide FEFR (*i.e.*: with no added thermolysin) within the concentration range investigated; 0 - 100 mg mL^{-1} suggesting that this short peptide does not self-assemble in water at pH 7. In contrast, when 0.3 mg mL^{-1} of themolysin enzyme was added self-supporting hydrogels were obtained depending on the initial tetra-peptide concentration, C_0, and the incubation time, t_{inc} (Figure 2). For $C_0 \leq 50$ mg mL^{-1} samples remained in the liquid state and no increase in viscosity was noted even after several days of incubation. Samples with $C_0 \sim 75$ mg ml^{-1} exhibited an increase in viscosity and formed soft gel that still flowed upon inversion after 13 hours. For $C_0 = 100$ mg mL^{-1} self-supporting hydrogels formed after 4 hrs. The macroscopic gelation time, t_{mgel} (time at which the sample stopped flowing upon inversion of the vial), was found to decrease with increasing initial tetra-peptide concentration.

Figure 2. Optical photographs for 25 after 48 hrs of incubation, 50 after 48 hrs of incubation, 75 after 13 hrs of incubation and 100 mg mL^{-1} sample after 2 hrs of incubation (left to right) showing their macroscopic appearance (solution / gel) after the addition of thermolysin (0.3 mg mL^{-1}).

The dynamics of the gelation process were followed using oscillatory rheology. In Figure 3 the evolution of the storage, G', and loss, G'', moduli for samples prepared with an initial concentration of tetra-peptide of 75 and 100 mg mL^{-1} are presented as a function of t_{inc}, the incubation time. For C_0 = 50 mg mL^{-1} (data not shown) G'' was always larger, or of the same order of magnitude, than G' during the 6 hrs probed confirming our visual observations that samples remain in a liquid state at low C_0 (data not shown). Different behaviour was observed when C_0 = 75 mg mL^{-1} (Figure 3). For t_{inc} < 195 min G'' was larger than G' indicating that the sample was in a liquid state, while for t_{inc} > 205 min G' was larger than G'' which is indicative of solid-like behaviour typical of physically entangled fibrillar gels. For C_0 = 100 mg mL^{-1} G' was found to increase faster and the cross-over point between G' and G'' was found at 60 min for this sample (Figure 3).

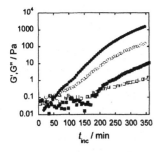

Figure 3. Evolution of the storage, G', (closed symbols) and loss, G'', (open symbols) moduli 75 mg mL^{-1} (square) and 100 mg mL^{-1} (circle) samples as a function of incubation time.

Sample compositions

MALDI-TOF MS and RP-HPLC were used to monitor qualitatively and quantitatively samples composition at discrete time points after the addition of thermolysin. Reference chromatographs of pure di-, tetra-, hexa- and octa-peptide sequences provided the range of retention times corresponding to the different species (10 – 13 min for di-peptide, 18 – 21 min for tetra-peptide, 21 – 23 min for hexa-peptide, 23 – 27 min for octa-peptide). The overall hydrophobicity of the peptides affects the elution time of the peptide species and the more hydrophobic the sequence the longer the retention time. A typical example of an RP-HPLC chromatograph obtained for a sample with an initial concentration of tetra-peptide of 100 mg mL^{-1} incubated for 32 hrs at room temperature and MALDI-TOF MS spectrum obtained for the same sample after 6 hrs incubation time are given in Figure 4.

The assignments were confirmed by a combination of elution times obtained from the RP-HPLC chromatographs of pure peptides, and MALDI-TOF MS results obtained by collecting the sample fractions corresponding to the different HPLC peaks. The complex peak shape observed by chromatography and the abundance of mass peaks in the MALDI-TOF spectrum indicate that a dynamic library of peptides with varied sequences were formed, reflecting the combinatorial nature of the reverse hydrolysis process. Results indicate that both the peptide length and quantities formed over time are dependant on C_0.

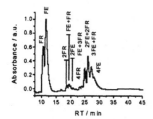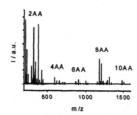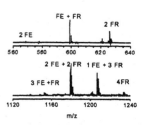

Figure 4. Left HPLC chromatograph with sequence assignment for the 100 mg mL^{-1} sample incubated for 32 hrs; middle MALDI-TOF mass spectrum after 6 hrs incubation times for the 100 mg mL^{-1} sample with sequence assignment of the peaks; and right expansion of the tetra- and octa-peptide regions of the MALDI-TOF mass spectrum.

The evolution of the relative percentages of di-, tetra-, hexa-, octa- and deca-peptides for samples with initial concentrations of 25, 75 and 100 mg mL^{-1} are presented in Figure 5 as a function of incubation time. For C_0 = 25 mg mL^{-1} (Figure 5) hydrolysis of the tetra-peptide occurs as soon as thermolysin is added and mainly di- and tetra-peptides are present for $t_{inc} \geq$ 10 hrs. Their concentrations $t_{inc} \geq$ 10 hrs are constant suggesting a chemical equilibrium between the hydrolysis and the reverse-hydrolysis reactions is established; the equilibrium composition of the solution being *ca.* 10 % tetra-peptides and 90 % di-peptides. Nevertheless it is interesting to note that trace (~ 3 %) of hexa-peptides is detected at 1 hrs incubation although they disappear with time. This suggests that some reverse-hydrolysis does indeed occur even at low C_0. In parallel to the disappearance of hexa-peptides, traces octa-peptide sequences appear after 2 hrs (ca. 2-3 %) suggesting that the hexa-peptides are unstable and further extended to octa-peptides by addition of di-peptide.

For C_0 = 50 mg mL^{-1} (data not shown) the tetra-peptide is hydrolysed immediately upon addition of thermolysin and after 30 min a similar sample composition as the 25 mg mL^{-1} sample is observed: 90 % di-peptides and 10 % tetra-peptides. From 30 min onwards longer sequences start to form. At t_{inc} = 2 hrs hexa- and octa-peptide are both present at the relative concentrations of 2.4 % and 0.8 % respectively. As t_{inc} increases further the quantity of hexa-peptides present in the sample is found to decrease until no hexa-peptides are detected when $t_{inc} \geq$ 6 hrs. This confirms that hexa-peptides are not stable species. The quantity of octa-peptides present in the sample is found to increase continuously over the incubation period probed. From 24 hrs onwards deca-peptides are also found to form and their concentration increases continuously during incubation. From t_{inc} = 4 hrs onwards the amount of di-peptides decreases at a constant rate suggesting that a steady state reverse-hydrolysis rate of reaction is achieved. Even after 96 hours no equilibrium is reached and a small but significant quantity of deca-peptides (ca. 5 %) is present in the sample. It should be noted that longer species are also likely to form in small quantities although they could not be quantified by HPLC due to their low concentration.

For C_0=75 mg mL^{-1}, similar kinetics to that observed for C_0=50 mg mL^{-1} is found. The tetra-peptide is rapidly hydrolysed into di-peptide and after 30 min of incubation onwards longer sequences were detected (hexa- and octa-peptides). As discussed previously the hexa-peptides content is found to decrease and disappears for t_{inc} > 6 hrs. This is concomitant with an increase in octa-peptide content and the appearance of deca-peptide sequences. It is interesting to note that from t_{inc} = 5 hrs onwards the amount of di- and tetra-peptides decrease at a constant rate

130

while the octa-peptides content increase at a constant rate. This suggests a steady state reverse-hydrolysis rate of reaction is achieved, and even after 72 hours no equilibrium is reached. For the $C_0 = 100$ mg mL^{-1} sample the tetra-peptide is again hydrolysed immediately after addition of thermolysin, and is followed by the formation of longer peptide sequences. In this case the process is significantly faster and after 2 hrs of incubation 7 % of octa-peptides are detected along with 78 %, 12 % and 3 % di- tetra- and hexa-peptides respectively. From 0.5 to 48 hrs of incubation a significant increase in octa-peptides (up to 35-40 %) and deca-peptides (up to 10 %) is observed. This increase is accompanied by a sharp decrease in di-peptides (down to 38 %) and a slight decrease in tetra-peptides (down to 50 %). For this sample hexa-peptides are again detected at the early stages and then disappear with increasing t_{inc}. From 48 hrs onward a slower increase in octa-peptides is observed and from 72 hours onwards the concentrations of the different peptides species become roughly constant suggesting that equilibrium is reached at this C_0. The composition of the system after 72 hrs is di- (40 %), tetra- (5%) octa- (40 %) and deca-peptides (15 %).

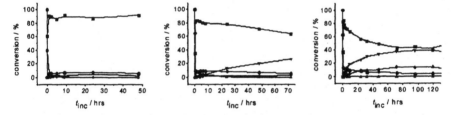

Figure 5. Evolution of the peptide bond content (conversion %) corresponding to each peptide sequence, di- (–!–), tetra- (–,–), hexa- (–7–), octa- (–B–) and deca-peptides (–Λ–), as a function of incubation time for (left to right) 25 mg mL^{-1}, 75 mg mL^{-1} and 100 mg mL^{-1} samples.

Nature of the fribillar network, oloigo-peptide secondary structure

ATR-FTIR was used to structurally characterise our samples and confirm the presence of β-sheet rich fibres. The ATR-FTIR spectra obtained for the 25, 75 and 100 mg mL^{-1} samples are presented in Figure 6 as a function of the incubation time. The absorption bands observed at 1148, 1200 and 1678 cm^{-1} are due to the presence of residual TFA in our original peptide samples. As the concentration of TFA is constant for a given sample, and proportional to C_0 (all the samples were prepared using the same batch of tetra-peptide), the band at 1148 cm^{-1} was used to normalise all ATR-FTIR spectra. The absorption bands at 1400 and 1555 cm^{-1} correspond to the amide II and amide III bands respectively. The intensities of these two absorption bands increase significantly for the $C_0 \geq 50$ mg mL^{-1} samples (Figure 6) due to the creation of amide bonds resulting from the synthesis of octa- and deca-peptides. In addition for these samples a strong band at 1622 cm^{-1}, and a weaker band at 1693 cm^{-1}, appear and grow with increasing t_{inc}. These bands have previously been assigned to peptides adopting an anti-parallel β-sheet arrangement.[7, 8] These results suggest that for $C_0 > 50$ mg mL^{-1} self-assembly of the peptide does occur and anti-parallel β-sheet rich fibrils form. This is in good agreement with our observation that an increase in viscosity occur ($C_0 = 50$ and 75 mg mL^{-1}) and subsequently a macroscopic gel is formed for $C_0 = 100$ mg mL^{-1}. For $C_0 = 25$ mg mL^{-1}, no fibrils were formed as suggested by the absence of absorption bands at 1622 and 1693 cm^{-1} even at early incubation times when a limited amount of octa-peptides were observed to form. This

131

indicates that the concentration of octa-peptides attained in this sample is probably below the critical self-assembling concentration.

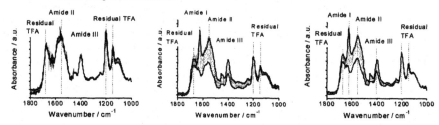

Figure 6. Time resolved FTIR spectra collected as a function of incubation time for the 25, 75 and 100 mg mL⁻¹ samples (left to right). Thick black lines correspond to $t = 0$ and 10 hrs.

CONCLUSIONS

In this paper we have investigated the possibility of using the protease thermolysin to trigger the reverse-hydrolysis and subsequent gelation of ionic-complementary peptides. Thermolysin was added at a fixed concentration (0.3 mg mL⁻¹) to aqueous solutions (0-100 mg mL⁻¹) of a short tetra-peptide (FEFR). The simultaneous analysis of MALDI-TOF MS and RP-HPLC chromatographs as a function of incubation time allowed the composition of the system to be elucidated and it was found that different peptide sequences formed depending on the initial concentration of tetra-peptides, C_0. For all samples the protease initially hydrolysed the tetra-peptides into the di-peptides FE and FR. Subsequently longer peptide sequences (hexa-, octa- and deca-peptides) were found to form through reverse-hydrolysis. At low C_0 ($\leq$ 50 mg mL⁻¹) these longer sequences were found to be unstable and disappeared with increasing incubation time. As C_0 was increased (> 50 mg mL⁻¹) the octa-peptide became the dominant product of the reverse-hydrolysis reaction. As the quantity of octa-peptide synthesised increased, self-assembly into anti-parallel β-sheet rich fibres was observed, as confirmed by ATR-FTIR. This novel method of forming ionic peptides, and subsequently hydrogels, will undoubtedly have significant impact in the synthesis of longer peptides and will also find important applications for the formation of scaffolds for in vivo tissue engineering, or for enzyme detection.

REFERENCES

1.   S. G. Zhang, C. Lockshin, R. Cook and A. Rich, Biopolymers 34, (1994)
2.   S. G. Zhang, T. Holmes, C. Lockshin and A. Rich, Proc. Natl. Acad. Sci. U. S. A. 90, (1993)
3.   A. Mohammed, A. F. Miller and A. Saiani, Macromol. Symp. 251, (2007)
4.   A. Saiani, A. Mohammed, H. Frielinghaus, R. Collins, N. Hodson, C. M. Kielty, M. J. Sherratt and A. F. Miller, Soft Matter 5, (2009)
5.   J. B. Guilbaud, E. Vey, S. Boothroyd, A. M. Smith, R. V. Ulijn, A. Saiani and A. F. Miller, submitted for publication to Langmuir (2010)
6.   K. Morihara and H. Tsuzuki, Eur. J. Biochem. 15, (1970)
7.   A. Barth, Biochim. Biophys. Acta-Bioenerg. 1767, (2007)
8.   A. Barth and C. Zscherp, Q. Rev. Biophys. 35, (2002)

Mater. Res. Soc. Symp. Proc. Vol. 1272 © 2010 Materials Research Society   1272-NN05-08

Using Peptide Hetero-Assembly to Trigger Physical Gelation and Cell Encapsulation

Andreina Parisi-Amon[1], Cheryl Wong Po Foo[2], Ji Seok Lee[2], Widya Mulyasasmita[1], and Sarah Heilshorn[2]
[1]Bioengineering and [2]Materials Science and Engineering, Stanford University, 476 Lomita Mall, Stanford, CA 94305, U.S.A.

ABSTRACT

Stem cell transplantation holds tremendous potential for the treatment of various trauma and diseases. However, the therapeutic efficacy is often limited by poor and unpredictable post-transplantation cell survival. While hydrogels are thought to be ideal scaffolds, the sol-gel phase transitions required for cell encapsulation within commercially available biomatrices such as collagen and Matrigel often rely on non-physiological environmental triggers (e.g., pH and temperature shifts), which are detrimental to cells. To address this limitation, we have designed a novel class of protein biomaterials: Mixing-Induced Two-Component Hydrogels (MITCH) that are recombinantly engineered to undergo gelation by hetero-assembly upon mixing at constant physiological conditions, thereby enabling simple, biocompatible cell encapsulation and transplantation protocols. Building upon bio-mimicry and precise molecular-level design principles, the resulting hydrogels have tunable viscoelasticity consistent with simple polymer physics considerations. MITCH are reproducible across cell-culture systems, supporting growth of human endothelial cells, rat mesenchymal stem cells, rat neural stem cells, and human adipose-derived stem cells. Additionally, MITCH promote the differentiation of neural progenitors into neuronal phenotypes, which adopt a 3D-branched morphology within the hydrogels.

INTRODUCTION

Physical hydrogels, characterized by transient crosslinks and shear thinning behavior, are ideal vehicles for minimally invasive cell transplantation. In recent years, research has been increasingly focused on developing cell-based therapies for a myriad of clinical needs, from spinal cord regeneration [1] to combating degenerative diseases such as Parkinson's [2] and multiple sclerosis [3]. However, this research has been met with many challenges. The success of cell transplantation therapies depends directly on the viability of the injected cells [2, 4, 5]. Unfortunately, many current cell injection techniques result in a substantial loss of transplanted cells [6,7]. A current goal is to create a scaffold that will contain the cells, protect them during the injection process, maintain them within the desired therapeutic location, and promote their growth and differentiation.

While hydrogels are seen as an ideal material for implantation due to their biomimetic nature and high water content, which requires introducing minimal levels of foreign material into the body and allows for high diffusivity of biomolecules [8], cell encapsulation is usually governed by external environmental triggers [9-13]. Examples of these triggers include temperature sweeps from 4 to 37 °C for collagen [14] and matrigel [15], pH shifts from ~2.5 to 7.4 for PuraMatrix [14] and cation concentration increases from 20 to 200 mM for alginate [14] and peptide amphiphiles [10]. Given that the materials are designed to be gels within the body, before transplantation the cells and accompanying proteins are often subjected to irreversibly damaging, non-physiological conditions [16]. In addition, these conditions are often difficult to reproducibly control, making them difficult to translate to clinical settings [16].

To combat these difficulties, we have harnessed protein-protein interactions between specific peptide domains to design Mixing-Induced, Two-Component Hydrogels (MITCH), where the two components contain separate peptide domains that associate simply upon mixing at constant physiological conditions (Figure 1). Recombinant protein engineering strategies were employed in the design and synthesis of the individual block copolymer components in MITCH [17]. The two association domains were selected by three primary design criteria. Firstly, the domains must be located intra-cellularly in their native context, so as not to stimulate undesired outside-in cell signaling pathways when incorporated in the cell-culture scaffold. Secondly, the domains must exhibit robust expression and folding when synthesized as tandem repeats in a recombinant host. Finally, the binding interaction of the association domains must be well-characterized, specific, and tunable to facilitate control over the crosslinking strengths that subsequently dictate the viscoelastic properties of the physical hydrogel. These criteria were satisfied by the WW domain family, which associate with proline-rich peptide ligands in a 1:1 stoichiometric ratio. Applying the hydrogel percolation theory to these chosen domains, we tailored two design variables (domain association energy and repeat frequency) to give MITCH tunable viscoelastic properties, making it possible to optimize the construct for encapsulation, proliferation, and differentiation of different cell systems.

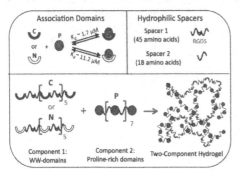

Figure 1: Schematic of MITCH. Two WW domains (C and N) bind to the same proline-rich peptide (P). The polymeric components are C7, N7, and P9. Component 1, C7 or N7, is composed of seven WW domains connected by RGD-containing hydrophilic spacers. Component 2, P9, is composed of nine proline-rich domains connected by hydrophilic spacers. A hydrogel is formed when component 1 (C7 or N7) is mixed with component 2 (P9).

EXPERIMENT

Recombinant protein engineering strategies were employed in the design and synthesis of the individual block copolymer components in MITCH. Full DNA sequences of the WW domain block copolymers (N3-7 and C3-7) and the corresponding proline-rich peptide chain (P3-9) were created from oligonucleotides (Integrated DNA Technologies, Coralville, IA) and incorporated into the pET-15b vector (Novagen, San Diego, CA) using standard cloning protocols. The BL21(DE3) *Escherichia coli* strain (Novagen, San Diego, CA) was used as the recombinant host, in which over-expression of the encoded protein was induced by 0.5 mM isopropyl β-D-1-thiogalactopyranoside (IPTG). Purification was achieved by affinity column chromatography, relying on the specific binding of an N-terminal polyhistidine tag to the Ni-NTA resin (Qiagen, Valencia, CA).

Biophysical characterization of the binding interactions between component 1 (C3 or N3) and component 2 (P3) was conducted by tryptophan fluorescence quenching and isothermal titration calorimetry. Fluorescence measurements (λ_{ex} = 295 nm, λ_{em} = 340 nm) were conducted on a SpectraMax Gemini EM spectrofluorometer (Molecular Devices), using 10^{-9} to 10^{-3} M of P3. Secondary structures of C3, N3, and P3 were probed by circular dichroism (CD), using an AVIV model 202-01 circular dichroism spectrometer equipped with a Peltier temperature-controlled cell holder (AVIV Biomedical, Lakewood, NJ). 0.0625 mg/ml of C3 or N3, and 0.125 mg/ml of P3 were used in the far-UV CD measurements, while 2.5 mg/ml of C3 or N3, and 5 mg/ml of P3 were used in the new-UV region. Circular dichroism experiments were done in buffer P (50 mM phosphate buffer, pH 7.5) at 4, 25, and 37 °C.

Viscoelastic properties were studied with micro-rheology, tracking the thermal motion of 0.2 μm fluorescent tracer particles (Molecular Probes). Unless otherwise noted, characterization was carried out in phosphate buffered saline (PBS). For each measurement, 10 μl of well-mixed sample was immediately pipetted in between a microscope slide and a cover slip, which were kept apart by a 120 μm thick SecureSeal™ Imaging Spacer (Grace Biolabs, Bend, OR). The Brownian trajectory of 20-50 beads were captured by a camera (iXON DV897 camera, ANDOR Technology, South Windor, CT) at ~30 Hz temporal resolution for ~15 seconds under a 40x oil objective fluorescence microscope. Image data was analyzed using IDL software version 7.0, following methods and macros developed by Crocker and Grier [18].

The effect of domain repeat frequency on material viscoelasticity investigated by embedding the fluorospheres within C3, C7, N3, N7, P3, and P9 solutions (7.5 wt%), as well as C3:P3, C7:P3, C3:P9, and C7:P9 mixtures (7.5 wt%). The concentration dependence of gelation behavior similarly characterized using 2.5 to 10 wt% of C7:P9 or N7:P9 hydrogels.

Shear-thinning behavior was studied by syringe injection, where C7:P9 and N7:P9 (10 wt%) hydrogels were manually ejected through a 26 gauge needle (5/8 inch length, ~0.42 mm diameter, BD, Franklin Lakes, NJ) The hydrogels were pre-equilibrated in the syringe for 60 minutes before shear-thinning, and particle tracking videos were captured prior to and at various times after injection.

Hydrogel formation in the presence of common extracellular molecules was examined by formulating the C7:P9 hydrogel (7.5 wt%) in PBS or Neurobasal-A cell culture medium, in the absence or presence of 5 wt% fetal calf serum (FCS).

Dissociated murine adult neural stem cells (NSC) were cultured as a single cell monolayer in polyornithine-laminin coated tissue culture plastic flasks using Neurobasal-A media supplemented with 1% glutamax, 2% B27 vitamin, 20 ng/ml epidermal growth factor (EGF) and 20 ng/ml fibroblast growth factor (FGF).

For NSC encapsulation, 5 μL of cells at 20×10^6 cells/mL were initially mixed with 5 μL of C7 or N7 solution. This mixture was then combined with 10 μL of P9 solution on a glass coverslip. Gel constructs were contained within Teflon washers and covered in media.

For NSC differentiation, cells were primed for 1 day in adult NSC growth media before changing to differentiation medium (low FGF concentration: 5 ng/ml) for 2 days and finally, replacing with Neurobasal-A media supplemented with only 1% glutamax and 2% B27 vitamin for 3 days.

Human adult adipose-derived stem cells (hASC) were cultured as monolayers on tissue culture plastic dishes using Dulbecco's Modified Eagle Medium (DMEM) plus glutamax, supplemented with 10% fetal bovine serum and 1% penicillin-streptomycin solution (v/v). Media was changed every three days. Trypsin (0.05%) was used to passage the cells.

For hASC encapsulation, 5 μL of cells at 5×10^6 cells/mL were initially mixed with 5 μL of C7 or N7 solution. This mixture was then combined with 10 μL of P9 solution on a glass coverslip. Gel constructs were contained within Pryex cloning cylinders and covered in media. For hASC differentiation, cells were primed for one day in hASC growth media (described above) before changing to osteogenic differentiation media, growth media supplemented with 250 μM ascorbic acid and 10 mm [beta]-glycerophosphate for a period of 10 days. Again media was changed every three days.

DISCUSSION

The creation of a hydrogel is dependent on the formation of a fully percolating polymeric network. The first step to achieving this goal was creating polymer strands of multiple WW or P domains. This was done through the incorporation of hydrophilic spacers to connect the individual domains. Spacer 1 connects the WW domains (either C or N) and was designed to be ~25 Å, about the length of a single WW domain as characterized by X-ray crystallography [19]. Spacer 2 connects the P domains and was designed such that its length is a non-integer fraction (2/5) of Spacer 1. This is to avoid the possibility of a single component 1 chain and a single component 2 chain interacting exclusively with each other and not taking part in the network formation.

The first tailored design variable was domain frequency. Components with varying domain frequencies were created by making C3, N3, P3, C7, N7, and P9 polymer strands, which as their names suggest are made of three connected C, N, or P domains; seven C or N domains; or nine P domains, respectively. Circular dichroism was preformed on the three-repeat polymer strands to confirm that fusion to the hydrophilic random coils did not affect proper folding of the association domains. Micro-rheology conducted on the C3:P3 versus C7:P9 gels clearly showed a difference between mixtures with differing domain frequencies. On the time scale of these experiments, all the mixtures except for C7:P9 behaved as viscous Newtonian fluids, seen by the fact that the log mean squared displacement (MSD) data scaled linearly with log time (τ) with a slope of 1. For the C7:P9 mixture, MSD data was found to be independent of time, consistent with the formation of a hydrogel network. This can be understood by considering the functionality (number of potential crosslinks per chain) of the different mixtures in the context of the mean-field percolation model of gelation. When the C and P polymers are mixed together, they begin creating a network through the association of C and P domains on adjacent chains, forming transient physical crosslinks. The average number of crosslinks per chain is found by multiplying the probability that a crosslink is formed (p) by the functionality of the chain (f). A stable hydrogel is formed when these crosslinks create a percolating network, which occurs when p is above the gel-point threshold (p_c). In the mean-field percolation model of gelation, p_c decreases as f increases [20]. In the above-mentioned experiment, the total number of potential crosslinks per gel was kept constant by making all gels 7.5% w/v protein. The differing levels of functionality made it possible for the C7:P9 mixture to form a gel while the other combinations (C3:P3, C7:P3, and C3:P9) continued to exhibit liquid-like behavior. An important control was that of confirming gelation due to domain association, as opposed to entanglement of the component chains. Rheological data showed that individual C7, N7, and P9 components are viscous Newtonian fluids at 7.5% w/v, while the C7:P9 and N7:P9 mixtures are viscoelastic gels at the exact same solution weight percent.

The second tailored design variable was the domain association energy. WW domains are protein units that mediate the interactions of proteins by binding proline-rich peptide motifs. As such, they are readily available in nature and have been well characterized. This makes it possible to find WW domains of different binding strengths, both by screening naturally occurring domains, as well as by computationally deriving new ones [21, 22]. Two domains were chosen: computationally derived CC43 (C) and a slight variation of wild-type Nedd4.3 (N) [21, 22]. These domains have been shown to bind to proline-rich PPxY peptides (P) with dissociation constants differing by an order of magnitude (1.7±0.1 µM for N and 11.2±1.2 µM for C) [22]. The modulation of binding affinity allows for control over the interaction energies between the domains of the two components (C and P or N and P). This level of control at the molecular level ultimately leads to control over mechanical properties at the macroscopic level, as the tighter binding between polymers formed of C and P components results in a greater number of crosslinks and formation of a more rigid gel in comparison to the N and P mixture. In micro-rheology experiments, the N7:P9 gel exhibited a higher plateau MSD value compared to the C7:P9 gel, implying that the latter gel is stiffer. These findings were corroborated by bulk oscillatory rheology data to determine the frequency-dependent storage (G') and loss (G") moduli of 10% w/v gels. For both gels the plateau storage modulus was larger than the loss modulus, indicating the formation of a gel. At room temperature, the more tightly binding C7:P9 gel was found to exhibit a storage modulus of ≈50 Pa, which is similar to 100% Matrigel. Under the same conditions, the more weakly binding N7:P9 mixture formed a gel with a storage modulus of ≈10 Pa, which is similar to 50% Matrigel [23].

The weak, transient crosslinks creating the hydrogel gel networks also provide the advantage of being shear-thinning and self-healing, thus making the gel injectable, an important quality for minimally-invasive cell transplant materials [6, 7, 12]. Micro-rheology data collected during self-healing, immediately following hand injections of fully formed hydrogels, confirmed the ability of both gels to self-heal, C7:P9 within 5 min and N7:P9 in 30 min. These results confirm the MITCH materials as candidates for cell-encapsulation and use as clinically injectable hydrogels whose molecular-level tuning abilities make it possible to modulate the kinetics of binding and self-healing.

The selection of specific molecular-recognition sites, natively found intra-cellularly, makes it possible for the MITCH system to be applicable across cell culture systems, as the tunable viscoelastic properties are unaffected by media components. This was confirmed using micro-rheological measurements, which showed that 7.5% w/v C7:P9 gels formed in either PBS or Neurobasal-A cell culture medium, either with or without 5% v/v fetal calf serum (FCS), exhibited the same viscoelastic properties. This ensured reproducibility of MITCH, even in the presence of additional biomolecules, is a key component to its use in a clinical setting, as well as across many cells systems. To this end, human endothelial cell, rat mesenchymal stem cells, rat neural stem cells, and human adipose-derived stem cells have been successfully cultured in 3D MITCH systems. Taking z-stack confocal images of the constructs 5-11 days post-encapsulation has shown the retention of the 3D nature of the culture system. Coupled with live (calcein-AM) and dead (ethidium-homodimer) cell staining, the images confirmed cell viability. The MITCH materials were able to support the proliferation and pluripotent state of rat neural stem cells in addition to their differentiation into glial and neuronal phenotypes. The differentiated cells adopted a 3D-branched morphology, with neurites extending over hundreds of microns. MITCH constructs are also currently being investigated for their ability to promote proliferation of human adipose-derived stem cells and their subsequent differentiation into osteogenic and adipogenic phenotypes.

CONCLUSIONS

Combining biomimetic design principles and hydrogel percolation theory, we designed and synthesized MITCH. These physical hydrogels display viscoelastic properties that are tunable via molecular-level engineering and reproducible across multiple cell-culture systems. In particular, we have shown its ability to support and promote growth and differentiation of neural stem cells into glial and neuronal phenotypes and promote the growth and viability of adipose-derived stem cells. Future work is aimed at further optimizing the system for 3D cell encapsulation, culture, and differentiation of neural and adipose-derived stem cells.

ACKNOWLEDGMENTS

We acknowledge funding from NIH DP2OD006477, NIH R01DK085720, the National Heart Foundation through the American Health Assistance Foundation, and the Stanford Bio-X Interdisciplinary Initiative. A.P.-A. and W.M. acknowledge a National Science Foundation Graduate Fellowship and a Stanford Graduate Fellowship, respectively.

REFERENCES

1. SM. Willerth , SE Sakiyama-Elbert , *Adv. Drug Delivery Rev.* **60**, 263 (2008).
2. A. Bjorklund, et al, *Brain Res.* **886**, 82 (2000).
3. C. Siatskas, CC. Bernand, *Current Molecular Medicine* **9**, 992 (2009).
4. JH. Kordower, et al, *Mov. Disord.* **13**, 383 (1998).
5. JH. Kordower, et a., *N. Engl. J. Med.* **332**, 1118 (995).
6. F. Cao, et al, *J. Tissue Eng. Regen. Med.* **1**, 465 (2007).
7. MA. Laflamme, et al, *Nat. Biotechnol.* **25**, 1015 (2007).
8. F. Brandl, F. Sommer, A. Goepferich, *Biomaterials* **28**, 134 (2007).
9. WA. Petka, et al, *Science* **281**, 389 (1998).
10. RM. Capito, et al, *Science* **319**, 1812 (2008).
11. L. Haines-Buterick, et al, *Poc. Natl. Acad. Sci. USA* **104**, 7791 (2007).
12. DJ. Pochan, et al, *J. Am. Chem. Soc.* **125**, 11802 (2003).
13. KL Niece, et al, *J. Am. Chem. Soc.* **125**, 7146 (2003).
14. BM. Gillette, et al, *Nat. Mater.* **7**, 636 (2008).
15. S. Wang, et al, *Tissue Eng. Part A* **14**, 227 (2008).
16. AI. Teixeira, JK. Duckworth, O. Hermanson, *Cell Res.* **17**, 56 (2007).
17. CTS. Wong Po Foo, et al, *Poc. Natl. Acad. Sci. USA* **106**, 22067 (2009).
18. JC. Crocker, DG Grier, *J. Colloid Interface Sci.* **179**, 298 (1996).
19. X. Huang, *Nat. Struct Biol.* **7**, 634 (2000).
20. PJ. Flory, *J. Am. Chem. Soc.* **63**, 3083 (1941).
21. V. KAnelis, D. Rotin, JD. Forman-Kay, *Nat. Struct. Biol.* **8**, 407 (2001).
22. WP. Russ, et al, *Nature* **437**, 579 (2005).
23. MH. Zaman, et al, *Poc. Natl. Acad. Sci. USA* **103**, 10889 (2006).

Hierarchical Self Assembly of Functional Materials— From Nanoscopic to Mesoscopic Length Scales

Mater. Res. Soc. Symp. Proc. Vol. 1272 © 2010 Materials Research Society   1272-OO01-05

Rolled-up helical nanobelts: from fabrication to swimming microrobots

Li Zhang and Bradley J. Nelson
Institute of Robotics and Intelligent Systems, ETH Zurich, Tannenstrasse 3, Zurich,
CH 8092, Switzerland

ABSTRACT

We present recent developments in rolled-up helical nanobelts in which helical structures are fabricated by the self-scrolling technique. Nanorobotic manipulation results show that these structures are highly flexible and mechanically stable. Inspired by the helical-shaped flagella of motile bacteria, such as E. coli, artificial bacterial flagella (ABFs) are a new type of swimming microrobot. Experimental investigation shows that the motion, force, and torque generated by an ABF can be precisely controlled using a low-strength, rotating magnetic field. These miniaturized helical swimming microrobots can be used as magnetically driven wireless manipulators for manipulation of microobjects in fluid and for target drug delivery.

INTRODUCTION

Microfabrication of helical structures is difficult for lithography-based techniques due to lithography's inherent 2-D patterning. Previously, a number of nanohelix fabrication methods have been developed based on "bottom-up" approaches [1-3]. Recentla, a strategy that combines "top-down" and "bottom-up" approaches for fabricating 3D micro-/nanostructures has been introduced [4-5]. This method is based on the coiling of strained 2D thin films to form 3D structures after the films detach from the substrate by selective etching, a type of self-assembly. Diverse 3D micro-/nanostructures have been achieved, such as tubes [4-6], helices [4, 7-8], micro-origami [9-11] and wrinkles [12-13]. It was found that rolled-up helical nanobelts can be designed with a specific geometrical shape, i.e., their diameter, chirality, pitch, helicity angle and length can be precisely controlled [14-15]. The as-fabricated helical nanobelts are highly flexible and retain a strong "memory" of their original shape [16-20]. Because of their interesting morphology and mechanical and electromagnetic properties, potential applications of these helical nanobelts include force sensors, chemical and biological sensors, inductors, and actuators. Inspired by the natural bacterial flagellum [21-23], we have developed helical swimming microrobots driven by a rotating magnetic field [24-26]. These miniaturized devices can be used as wireless manipulators for medical and biological applications in fluid environments, such as cell manipulation and removal of tissue. Due to its large surface to volume ratio, surface functionalized ABFs have the potential to sense and transmit inter- or intracellular information, and to perform targeted drug delivery.

EXPERIMENT

The fabrication of rolled-up helical nanobelts is illustrated schematically in Fig. 1. Bilayer or multi-layer thin films are deposited on a single crystal wafer, such as Si or GaAs, after the deposition of a sacrificial layer. The hetero-films are grown and patterned to a ribbon-like mesa by lithography. The ribbon is then detached from the substrate by selective wet etching of the sacrificial layer and forms a helical nanobelt to relieve internal stress. A more straightforward approach is to deposit hetero-films on a substrate with etching selectivity. Therefore, the coil structure can be obtained by directly etching the substrate [7]. Fabrication details are reported in Ref [14-17].

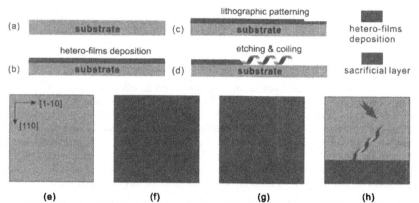

(e)   **(f)**   **(g)**   **(h)**

Figure 1. The flowchart for fabricating a rolled-up helical nanobelt on a (001)-oriented substrate. The red arrow indicates the scrolling direction of the nanobelt, i.e. [100].

A nanomanipulator and an AFM cantilever built in an SEM were used for the stretching test of individual helical nanobelts [17-20]. The manipulation processes were conducted as follows: a tungsten probe on the manipulator was first dipped into a silver tape to make its tip sticky. This probe was used to cut a helical nanobelt with one fixed end from the substrate and then to pick it up. Next, the free end of the spring was clamped onto an AFM cantilever using electron-beam-induced deposition (EBID). After the helical nanobelt was fixed between the probe and the AFM cantilever, as shown in figure 2, a tensile force was applied by moving the probe away from the AFM cantilever to investigate the spring constant of the helical nanobelt. Resonance frequency tests of a helical nanobelt were conducted using electrostatic actuation between a tungsten wire and the helical nanobelt [17].

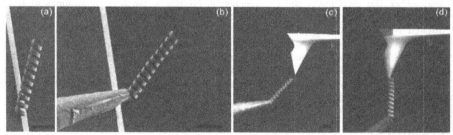

Figure 2. The nanorobotic manipulation of rolled-up helical nanobelts for characterizing flexibility and the spring constant. The scale bars are 5 μm.

An ABF consists of two parts, a helical nanobelt tail resembling a natural bacterial flagellum in both size and shape and a soft-magnetic metal head [24-25]. To actuate and control the swimming of the helical swimming microrobots wirelessly in water, three pairs of orthogonal electromagnetic coils were applied which are able to generate a uniform rotating magnetic field that exerts a torque on the head of the ABF, causing the ABF to rotate.

DISCUSSION

Controlled fabrication

Our fabrication results revealed that when the width of hetero-films is in the micrometer range or larger the scrolling direction for Si-based and GaAs-based strained films is determined by the smallest Young's modulus direction, e.g. <100> in the case of (001)-oriented substrate. Therefore, controlled fabrication of helices can be achieved, which means that the chirality, the helicity angle and the pitch of helices can be precisely controlled by the 2D ribbon-like mesa design. When the ribbon width is shrunk to the submicron and nanometer range, new effects from the edges of the ribbon cause the anomalous coiling of the nanohelices and strongly influence the shape of the rolled-up nanostructure. For example, experimental results showed that when the ribbon width is reduced to 300 nm for a 20 nm thick SiGe/Si bilayer, the effect of stress relaxation at the edges of the ribbon, rather than the Young's modulus, dominate the scrolling process (see Fig. 3). Stress relaxation leads to a uniaxial strain component along the long axis of the ribbon. Based on this anomalous coiling principle, nanocoils with small helical angles and pitches have been achieved, which exceeds the 45° limitation dictated by the preferred <100> scrolling direction for micrometer scale helices [14-15]. Depositing a metal layer on the bilayers can also be used to tailor the shape of the nanostructure. Our new findings of the anomalous coiling principle have provided additional design and fabrication opportunities to create 3D nanostructures.

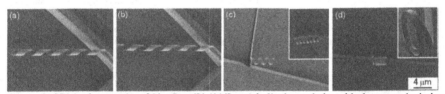

Figure 3. SEM micrographs of $Si_{0.6}Ge_{0.4}$/Si (11/8 nm) helical nanobelts with the same deviation of 5° from <110>. The ribbon width in (a) to (d) is 1.30 μm, 1.20 μm, 0.40 μm and 0.30 μm, respectively. All the images have the same magnification. Inset in (c) shows an SEM micrograph of an anomalous coiled SiGe/Si/Cr helical nanobelt with small pitch. Inset in (d) shows an SEM micrograph of a Si/Cr microspiral.

Mechanical properties

Tensile tests were performed for SiGe/Si/Cr and InGaAs/GaAs helical nanobelts and show that they are highly elastic and behave like micro-springs (see Fig. 4) with spring constants in a range of 3-20 pN/nm [17-20]. By tuning design parameters, i.e. the number of turns, the thickness of the wall, the width and orientation of the stripe, the diameter and the pitch of a helical spring, a desired stiffness can be obtained through simulation. Resonance tests of the helical nanobelts, both in the bending mode and in the axial displacement mode, indicated that the mechanical properties of the bilayer material do not change over 10^6 loading cycles [17].

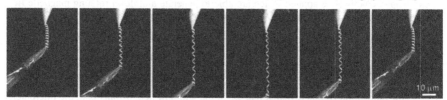

Figure 4. Spring constant characterization of a SiGe/Si/Cr helical nanobelt. According to the linear range of the elongation-load curve, the spring constant of the helical nanobelts is 0.003 N/m [18]. All SEM micrographs have the same scale bar.

Swimming behavior of ABFs

In 1973, it was discovered that some bacteria such as E. coli swim in liquid by turning the base of a flagellum or a bundle of flagella [21]. Unlike E. coli, which uses 45 nm diameter molecular motors for actuation [22], the ABF was actuated by a uniform rotating magnetic field. A magnetic torque is applied on the soft magnetic head to rotate the ABF along its helical axis and to steer it in liquid. Due to the helical nanobelt tail, the ABF is self-propelled somewhat similar to a corkscrew motion [23]. The results showed that these helical swimming microrobots can be steered in 3-D with micrometer positioning precision as shown in Fig. 5 [24]. Swimming tests of ABFs showed a linear relationship between the frequency of the applied field and the translational velocity when the frequency is lower than the step-out frequency of the ABF [25]. Moreover, the propulsive force and the applied torque can be estimated and tuned over a large range by changing the input field once the propulsion matrix is experimentally determined [25]. We calculated the maximum force and the applied torque generated by a 38 μm long ABF [25]. It was found that with a 2 mT magnetic field a maximum force of 3.0 pN is sufficient to move a microscale object such as a blood cell. The maximum torque is 43 nN·nm, which is ca. two orders of magnitude larger than that of the molecular motor in E. coli [22].

Figure 5: A 38 μm long ABF swims towards a target (dark dot) controlled by a rotating magnetic field [24].

Micromanipulation using ABFs

Self-propelled ABFs can provide a versatile tool for manipulating cellular or sub-cellular objects [26]. In principle, two different manipulation methods could be applied: (i) Contact manipulation, in which ABFs contact the microobjects and manipulate them directly. (ii) Noncontact manipulation, in which ABFs manipulate the microobjects by generating a controlled fluid flow. Contact manipulation, such as pushing and rotating (see Fig. 6), were discussed earlier [24]. For a manipulation task such as drilling biological membranes and drug delivery into a cell, the small cross-sectional area of the ABF has the advantage of applying a relatively large pressure. For noncontact manipulation, the fluid flow generated is able to pump microobjects to one side of the ABF simultaneously, once the boundary condition of the swimming ABF is inhomogeneous around the ABF [26].

Figure 6: Two 6 μm diameter polystyrene (PS) microspheres are rotated by a 29 μm long ABF [24].

CONCLUSIONS

In summary, the self-scrolling technology is promising for controlled fabrication of helical nanobelts, which have potential applications in MEMS/NEMS devices and wirelessly controlled swimming microrobots. Once functionalized, ABFs have the potential to perform as sensors for inter- or intracellular information sensing and to perform targeted delivery of energy (e.g., inductive heating) and chemical and biological substances. ABFs also have the potential to be used as in vivo medical micro/nanorobots. However, biocompatibility of the materials, tracking, and navigation of the robots in a dynamic fluidic environment remain a challenge.

ACKNOWLEDGMENTS

The authors thank the FIRST lab and EMEZ at ETH Zurich and the Laboratory for Micro- and Nanotechnology at the Paul Scherrer Institute for microfabrication and characterization support. The authors are grateful to Prof. Detlev Gruetzmacher (RWTH Aachen), Prof. Lixin Dong (Michigan State University), Prof. Jake J. Abbott (University of Utah), Dr. Dominik Bell (ETH Zurich), Dr. Bradley E. Kratochvil (ETH Zurich) and Kathrin E. Peyer (ETH Zurich) for their contributions. Funding for this research was partially provided by the Swiss National Science Foundation (SNSF).

REFERENCES

1. S. Amelinckx, X. B. Zhang, D. Bernaerts, X. F. Zhang, V. Ivanov and J. B. Nagy, Science, **265**, 635 (1994).
2. P. X. Gao, Y. Ding, W. J. Mai, W. L. Hughes, C. S. Lao and Z. L. Wang, Science, **309**, 1700 (2005)
3. B. A. Korgel, Science, **309**, 1683 (2005).
4. V. Y. Prinz, V. A. Seleznev, A. K. Gutakovsky, A. V. Chehovskiy, V. V. Preobrazhenskii, M. A. Putyato and T. A. Gavrilova, Physica E, **6**, 828 (2000).
5. O. G. Schmidt and K. Eberl, Nature, **410**, 168 (2001).

6. O. G. Schmidt, N. Schmarje, C. Deneke, C. Muller and N. Y. Jin-Phillipp, Adv. Mater., **13**, 756 (2001).
7. S. V. Golod, V. Y. Prinz, V. I. Mashanov and A. K. Gutakovsky, Semicond. Sci. Tech., **16**, 181 (2001).
8. L. Zhang, S. V. Golod, E. Deckardt, V. Prinz and D. Grutzmacher, Physica E, **23**, 280-284 (2004).
9. P. O. Vaccaro, K. Kubota and T. Aida, Appl. Phys. Lett., **78**, 2852 (2001).
10. W. J. Arora, A. J. Nichol, H. I. Smith and G. Barbastathis, Appl. Phys. Lett., **88**, 053108 (2006).
11. T. G. Leong, C. L. Randall, B. R. Benson, N. Bassik, G. M. Stern and D. H. Gracias, Proc. Natl. Acad. Sci. U. S. A. , **106**, 703 (2009).
12. Y. F. Mei, D. J. Thurmer, F. Cavallo, S. Kiravittaya and O. G. Schmidt, Adv. Mater., **19**, 2124 (2007).
13. A. Malachias, Y. F. Mei, R. K. Annabattula, C. Deneke, P. R. Onck and O. G. Schmidt, ACS Nano, **2**, 1715 (2008).
14. L. Zhang, E. Deckhardt, A. Weber, C. Schonenberger and D. Grutzmacher, Nanotechnology, **16**, 655 (2005).
15. L. Zhang, E. Ruh, D. Grutzmacher, L. X. Dong, D. J. Bell, B. J. Nelson and C. Schonenberger, Nano Lett., **6**, 1311 (2006).
16. L. Zhang, L. X. Dong, D. J. Bell, B. J. Nelson, C. Schoenenberger and D. Gruetzmacher, Microelectron. Eng., **83**, 1237 (2006).
17. D. J. Bell, L. X. Dong, B. J. Nelson, M. Golling, L. Zhang and D. Grutzmacher, Nano Lett., **6**, 725 (2006).
18. D. J. Bell, Y. Sun, L. Zhang, L. X. Dong, B. J. Nelson, D. Grutzmacher, Sensor Actuat. a-Phys. **130**, 54 (2006).
19. L. Zhang, L. X. Dong and B. J. Nelson, Appl. Phys. Lett., **92**, 143110 (2008).
20. L. X. Dong, L. Zhang, D. J. Bell, D. Grutzmacher and B. J. Nelson, Journal of Physics: Conference Series, **61**, 257 (2007).
21. H. C. Berg and R. A. Anderson, Nature, **245**, 380 (1973).
22. H. C. Berg, *E. coli in motion*, Springer-Verlag, New York, 2004.
23. E. M. Purcell, Am. J. Phys., **45**, 3 (1977).
24. L. Zhang, J. J. Abbott, L. X. Dong, B. E. Kratochvil, D. Bell and B. J. Nelson, Appl. Phys. Lett., **94**, 064107 (2009).
25. L. Zhang, J. J. Abbott, L. X. Dong, K. E. Peyer, B. E. Kratochvil, H. X. Zhang, C. Bergeles and B. J. Nelson, Nano Lett., **9**, 3663 (2009).
26. L. Zhang, K. E. Peyer and B. J. Nelson, submitted.

Mater. Res. Soc. Symp. Proc. Vol. 1272 © 2010 Materials Research Society

Polymer Tubes by Rolling of Polymer Bilayers

Kamlesh Kumar[1,2], Bhanu Nandan[1], Valeriy Luchnikov[3], Svetlana Zakharchenko[1], Leonid Ionov[1], Manfred Stamm[1]

[1] Leibniz Institute of Polymer Research Dresden
Hohe Str 6, 01069 Dresden

[2] present address: Oregon Health & Science University
3181 SW Sam Jackson Park Rd. Portland, USA

[3] Institut de Science des Materiaux de Mulhouse, LRC 7228
CNRS & Universite de Haute Alsace, 15, rue Jean Starcky,
Mulhouse 68057, France

ABSTRACT

Polymer micro- and nanotubes are of growing interest for design of microfluidic devices, chromatography, biotechnology, medicine chemical sensors, etc. One approach for the design of tubes is based on use of self-rolling thin films. Here we overview our recent progress in the fabrication polymeric self-rolling tube.

INTRODUCTION

Polymer micro- and nanotubes have been demonstrated to possess remarkable applications in various fields such as microfluidic devices [1], chromatography, biotechnology [2], medicine [3-4] and chemical sensors [5]. Among different methods for preparation of tubes, stress-induced self-rolling of thin films deserved a particular interest. This method was originally developed for design of inorganic (metal and metal oxide) tubes [6-7]. Recently, we applied this approach for fabrication of polymeric tubes. This paper overviews our recent developments in this field.

DESIGN OF SELF-ROLLING TUBES

Our approach is based on use of thin polymeric bilayers deposited on a substrate[8-9]. The rolling of tube is provided by the bending moment due to swelling properties of chemically dissimilar polymers in selective solvents (Figure 1). The lateral force, which creates the bending moment, arises in response to an unequal change in the specific volume of the components of the film. This principle was used for fabrication of polystyrene (PS)/poly (4-vinyl pyridine) (P4VP) film. The polymer bilayer was produced by consecutive deposition of PS and P4VP, from toluene and chloroform solutions, respectively. The tube formation proceeds from an opening in the film made by photolithography or by mechanical scratching followed by immersion of patterned sample in dodecylbenzene sulfonic acid (DBSA) solution or in acidic aqueous environment. DBSA forms supramolecular complexes with pyridine rings of P4VP and increases the specific volume of the polymer. Since the solution is neutral to PS layer, bilayer film develops strain due to unequal swelling of polymers in solution of DBSA and hence the film bends and scrolls in order to minimize its free energy and form tubes. Using layers with two-dimensional gradient of thickness [10], we thoroughly investigated process on tube formation with respect to acidity of the solution and UV dose. It was found that rate of rolling increased with the acidity of the solution. Tube diameter and rate of rolling decreased with the increase of the UV exposure time. Moreover, increase of thickness of PS results in increase of the diameter of tube.

Figure 1. Scheme of fabrication of polymeric microtubes with using polymeric bilayers.

SELF-ROLLING TUBES WITH PATTERNED INNER WALL

Possibility to functionalize the hidden walls of the tubes is one of the major advantages of the self-rolling approach[11]. One can modify the surface of the film prior to rolling by magnetron sputtering of metal and upon rolling, tube and toroids with metallized inner surface could be obtained. Using this approach, we have fabricated polymeric self-rolling tubes with the gold current conductive stripes on its inner walls (Figure 2).

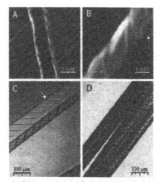

Figure 2 AFM and optical microphotographs of tubes with patterned inner surfaces. Reprinted with permission of ref. [11], copyright 2007 Wiley InterScience.

HYBRIDE TUBES

We also adopted the self-rolling phenomena of polymer layers to produce silica and silica/metal hybrid tube[12-13]. A thin film of PDMS was rolled with the P4VP/PS bilayer resulting in the formation of PDMS/P4VP/PS trilayer polymer tube. These polymer tubes were then pyrolyzed in an oxidative environment that converted PDMS into silica and simultaneously removed the organic part of the tube (Figure 3). Furthermore, it was shown that by depositing a thin film of metal along with the PDMS layer, it is possible to fabricate silica/metal hybrid tubes. The main advantage of this approach is the possibility of the formation of ceramic tubes of very high aspect ratio using a preceramic procedure. Furthermore, silica can be combined with any metal or combination of metals for fabricating silica/metal hybrid tubes.

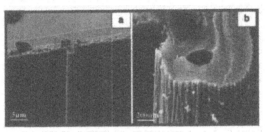

Figure 3. SEM microphotography of silica tube with open end (a) at lower magnification (b) at higher magnification. Reprinted with permission of ref.[12], copyright 2009 American Chemical Society.

SELF-ROLLING TOROIDS

Furthermore, polymer micro-toroids and triangles can be also fabricated using self-rolling approach of PS/P4VP layer[14]. The equilibrium dimensions of toroid are determined by the balance of the bending and the stretching energies of the film. The width of the rolled-up bilayer is larger for the films with higher values of the bending modulus and smaller values of the effective stretching modulus. Toroids can also be used for the study of the behaviour of quantum dots or cells in confined space.

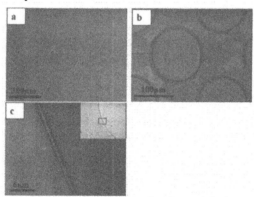

Figure 4. Optical micrographs of toroidal tubes.

APPLICATION OF SELF-ROLLING TUBES

The research on the self-rolled polymer tube can be extended by exploring these tubes as highly sensitive chemical or biological sensors. It was found during this work that polyelectrolyte molecule (conductive polymers) could be captured during the flow of polyelectrolyte solution in self-rolled tube. Moreover, these polyelectrolyte molecules could further be stretched in the confined space of polymer tube. The hydrodynamic forces can be used for stretching of the molecule and then it could be deposited between two gold electrodes. The deposited molecule will work as a bridge between two electrodes. Alternatively, one can also stretch the molecule or make thin film of conductive molecules between two gold electrodes before rolling of polymer layer and later it could be transferred to the interior of the tube during rolling of the surface. Outer stripes could be used as electrodes to measure the conductivity. This kind of device will have potential application in chemical or biological sensor since when some conductive moiety comes in contact with bridge molecule, the ionic current change. Hence, such sensors could be used to detect the conductive moiety.

Furthermore, we tested the applicability of self-rolling tubes as micro-resistors. In order to demonstrate this, we performed electrolysis reaction inside the self-rolled polymer. The tube was connected to current source via gold electrodes prepared as shown in Figure 5. After applying of current, a bubble consisting of oxygen and hydrogen was formed due to electrolytic decomposition of water.

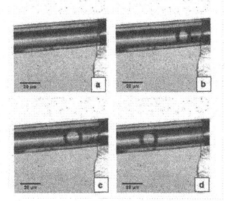

Figure 5. Generation of micron size bubble inside the polymer tube by electrolysis (a) stage before electrolysis, (b,c and d) are the consequent figures after bubble generation.

Recently, we also designed partially biodegradable thermo-magneto-sensitive self-rolling microtubes based polycaprolactone-poly(N isopropylacrylamide) bilayer which are able to reversibly roll-unroll as well as encapsulate-release microparticles in aqueous environment in response to temperature changes. Moreover, we demonstrated possibility to manipulate the tubes using magnetic field [15].

CONCLUSIONS

Self-rolling of polymer bilayers is a very convenient approach for interfacing the interior of microtubes with external electrical circuits and it can be used in particular for creating devices like micro-bubble generators exploiting electrolytic decomposition of fluids. In the present studies we have fabricated self-rolling tubes with inner diameter ranging from 1μm up to millimetres and with wall thickness ranging from 50nm up to several microns. On the other hand, we have already produced tubes with inner diameter below 100nm. The largest diameter is limited by the collapse of the tube upon drying and therefore also depends on wall thickness. In future, we plan to investigate mechanical properties of the polymeric self-rolling tubes, which are expected to be much softer than the metallic ones. We believe that self-rolled polymeric tubes could be of interest for microbubble generation inside the polymer tube as well as capture and release of particles and cells.

ACKNOWLEDGMENTS
The authors are thankful to Leibniz Institute of Polymer Research Dresden and DFG for financial support (grant IO 68/1-1).

REFERENCES
1. Bruzewicz, D. A.; McGuigan, A. P.; Whitesides, G. M. *Lab on a Chip* **2008**, 8, 663-671.
2. Martin, C. R. *Science* **1994**, 266, (5193), 1961-1966.
3. Andersson, H.; van den Berg, A. *Sensors and Actuators B-Chemical* **2003**, 92, 315-325.
4. McAllister, D. V.; Wang, P. M.; Davis, S. P.; Park, J. H.; Canatella, P. J.; Allen, M. G.; Prausnitz, M. R. *Proceedings of the National Academy of Sciences of the United States of America* **2003**, 100, (24), 13755-13760.
5. Huber, D. L.; Manginell, R. P.; Samara, M. A.; Kim, B. I.; Bunker, B. C. *Science* **2003**, 301, (5631), 352-354.
6. Vorob'ev, A. B.; Prinz, V. Y. *Semiconductor Science and Technology* **2002**, 17, 614-616.
7. Golod, S. V.; Prinz, V. Y.; Mashanov, V. I.; Gutakovsky, A. K. *Semiconductor Science and Technology* **2001**, 16, (3), 181-185.
8. Luchnikov, V.; Stamm, M. *Physica E-Low-Dimensional Systems & Nanostructures* **2007**, 37, (1-2), 236-240.
9. Luchnikov, V.; Stamm, M.; Akhmadaliev, C.; Bischoff, L.; Schmidt, B. *Journal of Micromechanics and Microengineering* **2006**, 16, (8), 1602-1605.
10. Kumar, K.; Luchnikov, V.; Nandan, B.; Senkovskyy, V.; Stamm, M. *European Polymer Journal* **2008**, 44, (12), 4115-4121.
11. Luchnikov, V.; Sydorenko, O.; Stamm, M. *Advanced Materials* **2005**, 17, (9), 1177-+.
12. Kumar, K.; Nandan, B.; Luchnikov, V.; Simon, F.; Vyalikh, A.; Scheler, U.; Stamm, M. *Chemistry of Materials* **2009**, 21, (18), 4282-4287.
13. Kumar, K.; Nandan, B.; Luchnikov, V.; Gowd, E. B.; Stamm, M. *Langmuir* **2009**, 25, (13), 7667-7674.
14. Luchnikov, V.; Kumar, K.; Stamm, M. *Journal of Micromechanics and Microengineering* **2008**, 18, (3).
15. Zakharchenko, S.; Puretskiy, N.; Stoychev, G.; Stamm, M.; Ionov, L. *Soft Matter* **2010**, DOI: 10.1039/c0sm00088d.

Mater. Res. Soc. Symp. Proc. Vol. 1272 © 2010 Materials Research Society   1272-OO02-07

Capillary And Magnetic Forces For Microscale Self-Assembled Systems

Christopher J. Morris[1], Kate E. Laflin[2], Brian Isaacson[1], Michael Grapes[1], and David H. Gracias[2]

[1]U.S. Army Research Laboratory, Adelphi, MD, 20783, USA

[2]Johns Hopkins University, Baltimore, MD, 21218, USA

ABSTRACT

Self-assembly is a promising technique to overcome fundamental limitations with integrating, packaging, and generally handling individual electronic-related components with characteristic lengths significantly smaller than 1 mm. Here we briefly summarize the use of capillary and magnetic forces to realize two example microscale systems. In the first example, we use capillary forces from a low melting point solder alloy to integrate 500 µm square, 100 µm thick silicon chips with thermally and chemically sensitive metal-polymer hinge actuators, for potential medical applications. The second example demonstrates a path towards self-assembling 3-D silicon circuits formed out of 280 µm sized building blocks, utilizing both capillary forces from a low melting point solder alloy and magnetic forces from integrated, permanent magnets. In the latter example, the utilization of magnetic forces combined with capillary forces improved the assembly yield to 7.8% over 0.1% achieved previously with capillary forces alone.

INTRODUCTION

Fluidic self-assembly is emerging as an attractive method to heterogeneously integrate devices in two and three dimensional configurations [1]. The technique typically involves micrometer-scale parts suspended in fluid, and a set of attractive forces to direct those parts to desired binding sites. Capillary forces are one type of force arising from submerged, immiscible fluids, and are useful to assemble parts with incompatible substrate materials [2-9], as well to form new, three-dimensional material and device structures [10-15]. Magnetic forces are also useful to assemble parts at magnetized template sites [16-18], to manipulate ferromagnetic parts by external fields [19], and to assemble magnetized parts with one another [20]. The manufacture of microsystems by assembling pre-microfabricated parts in different ways may enable many new applications in sensing, microrobotics, and even high performance computing, in part because the customization of wafer-scale manufacturing processes for every limited-use application may no longer be necessary.

We present methods to employ both short-range capillary forces, and the longer range attraction of magnetic forces, to realize two example systems. First, as a specific example of integration between incompatible materials, micrometer scale silicon chips standing in for commercially available microelectronic devices are integrated with self-actuating, metallic hinge actuators. These actuators have been used to create free-standing, self-folding three-dimensional polyhedra [21], [22], as well as microgrippers for microsurgery or other pick-and-place operations [23, 24]. These hinges are powered by metallic bilayers with intrinsic stress, stabilized by a third polymer layer until actuation is desired. This polymer layer can respond to specific thermal and chemical inputs by softening, and allowing the metallic bilayer to bend. In

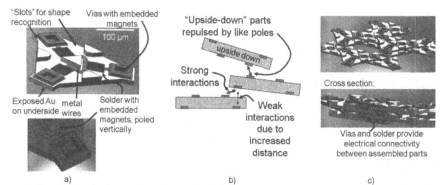

Fig. 1. Schematic showing capillary force based self-assembly using a molten alloy to form three dimensional electrical connections. a) Part features labeled. b) Possible magnetic interactions from pairs of poled, permanent magnets. c) Intended assembly and cross section showing electrical path between one part and the next through interlocking via and solder bump connections.

our process capillary forces from a low melting point alloy direct the assembly of silicon chips with these actuators prior to their release.

The second example system utilizes both capillary and magnetic forces, and is depicted in Fig. 1. Figure 1a) shows a triangular part containing several features designed to insure assembly with other parts to form a three dimensional, electrically-connected network. A slotted geometry and solder alloy deposited on the interior region of the slot allow complimentary shapes to insure correct orientation upon assembly. The geometry of the slot and location of the solder alloy ideally prevent any overlap of the alloy on one part with the alloy on another part encountered in an "upside down" orientation. Electrical vias near the exterior of the slot provide alloy-wetable binding sites on the underside, so that upon assembly in the correct orientation, the alloy and binding site overlap and a capillary bond forms. The interconnect metal, which connects vias, alloy bumps and any embedded circuitry is not wetable by the alloy, to prevent incorrect alloy binding. Furthermore, embedded permanent magnets at the underside of the via surface and at the top of the solder-covered surface help direct proper assembly. Figure 1b) shows how the embedded magnets interact to guide the intended assembly of the parts. Figure 1c) shows 3-D self-assembled structures, and a cross sectional detail of the 3-D electrical interconnects. These parts fabricated in a manner which is intended to be entirely compatible as a post process to any CMOS or wafer-scale packaged MEMS device, and which does not involve temperatures greater than those associated with photolithography steps.

EXPERIMENTAL DETAILS

Self-Assembly of Silicon Chips With Hinge Actuators

To fabricate bimetallic hinge actuator devices, we followed methods similar to those in [23, 24], and shown in Fig. 2. First, a sacrificial layer of copper (250 nm), a thin film layer of chrome (77 nm), and a layer of gold (300 nm) were thermally evaporated onto a silicon wafer as shown

154

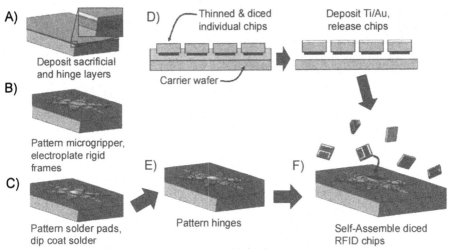

A) Deposit sacrificial and hinge layers

B) Pattern microgripper, electroplate rigid frames

C) Pattern solder pads, dip coat solder

D) Thinned & diced individual chips — Carrier wafer — Deposit Ti/Au, release chips

E) Pattern hinges

F) Self-Assemble diced RFID chips

Fig. 2: Microgripper fabrication process

in Fig. 2A, and the Cr and Au layers were patterned by lift-off in acetone. Photolithography defined rigid, "non-hinge" areas, and these areas were filled with a 4 μm thick layer of electroplated gold (Fig. 2B). Another photolithography step defined regions for solder binding sites at the center of each microgripper, and a 1 μm layer of electroplated copper provided a wetting layer for the solder. The entire wafer was dipped in a solder alloy with a melting point of 47°C (Alloy-117, Indium Corporation, Utica, NY) to create small solder pads at the center of each microgripper as shown in Fig. 2C. Figure 2E depicts the 3 μm thick Shipley 1827 photoresist (Microchem) patterned over the top of each hinge. This polymer prevented hinges from bending until after microgripper release, when applied heat would soften the polymer and allow each hinge to bend.

Meanwhile, silicon wafer fragments were thinned to a thickness of 100 μm and diced into 500 μm square chips while bonded to a carrier wafer using a wax (Crystalbond 509), shown in Fig. 2D. A sputtered 20 nm Ti and 200 nm Au layer defined solder binding sites, after which the chips were released by dissolving the wax. The chips were then self-assembled onto the solder pads of the microgrippers in 60°C water, with acetic acid added so the pH was approximately 2. Gently swirling a sealed jar containing the water, acid, chips, and substrate helped agitate the parts relative to the substrate, and capillary forces from the molten solder alloy captured and aligned chips into place on the microgrippers (Fig. 2F).

Finally, the microgripper/chip assemblies were released to become freestanding devices, after dissolving the sacrificial copper layer in ferric chloride.

Self-Assembly of 3D Circuit Elements

To fabricate the parts in Fig. 1, we started with a silicon-on-insulator (SOI) wafer having a 20 μm device layer as shown in Fig. 3A). We defined slotted features with a 13 μm deep reactive ion etch (DRIE), and 40 μm wide vias with another 7 μm DRIE step in Fig. 3B). Figure 3C) shows the process to partially fill the vias, first with a 4 μm thick electroplated

155

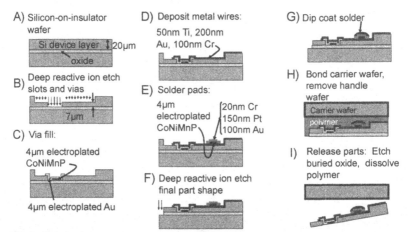

A) Silicon-on-insulator wafer

Si device layer 20µm

oxide

B) Deep reactive ion etch slots and vias

7µm

C) Via fill:

4µm electroplated CoNiMnP

4µm electroplated Au

D) Deposit metal wires: 50nm Ti, 200nm Au, 100nm Cr

E) Solder pads:

4µm electroplated CoNiMnP

20nm Cr
150nm Pt
100nm Au

F) Deep reactive ion etch final part shape

G) Dip coat solder

H) Bond carrier wafer, remove handle wafer

Carrier wafer

polymer

I) Release parts: Etch buried oxide, dissolve polymer

Fig. 3. Fabrication process for parts shown in Fig. 1 (for simplicity, only half of one part is shown in each cross sectional view).

CoNiMnP permanently magnetic material, followed by a 4 µm layer of electroplated Au. The magnetic material was electroplated from a bath formulated based on past reports [25-28], although this particular bath was further optimized following a series of tests detailed in an upcoming paper [29]. We measured the magnetic hysteresis of this material with other test wafers using a vibrating sample magnetometer. The Au was electroplated in a commercially-available bath (Technigold 25, Technic, Inc., Pawtucket, RI) at 49°C, 3 mA/cm^2, 1:10 duty cycle (1 ms on, 10 ms off). After photoresist removal, a Au etch solution (GE-8110, Transene, Danvers, MA) removed the seed layer. The metal layers shown in Fig. 3D) included a sputtered interconnect multilayer of 50 nm Ti, 200 nm Au, and 100 nm Cr, patterned by lift-off in acetone. This step was followed by the patterning and electroplating of another set of CoNiMnP magnetic features in Fig. 3E), and an evaporated multilayer consisting of 20 nm Cr, 150 nm Pt, and 100 nm Au, also patterned by lift-off in acetone, covered these magnets, . The final part outline was defined by a silicon DRIE step and an oxide RIE step depicted in Fig. 3F).

A eutectic Bi-Sn-Pb-In-Cd alloy was deposited as shown in Fig. 3G). First, a sputtered 50 nm of Au followed by 50 nm of Cu insured that the alloy would wet the entire substrate. We dipped the wafer through a 160°C ethylene glycol/solder interface, and the Au and Cu quickly dissolved allowing the alloy to react with the remaining patterned Pt layer. Upon removal of the substrate through this same interface, the alloy dewetted from other areas leaving the alloy only at the prescribed locations.

The final steps included Fig. 3H), where a carrier wafer was bonded to the device layer using a spin-on polymer (ProTEK A2, Brewer Sciences, Inc., Rolla, MO) and a wafer bonding tool at 1 Torr, 150°C, and 1 atm applied pressure. We then removed all but approximately 30 µm of the handle layer by lapping, and a XeF$_2$ etch removed the remaining silicon. Before releasing the parts, the wafer was placed in an 800 kA/m magnetic field to pole the magnets perpendicularly to the wafer plane. Finally, we etched the exposed buried oxide in 49% hydrofluoric acid, and dissolved ProTEK in a dodecene-based solvent (ProTEK Remover 200, Brewer Sciences). The

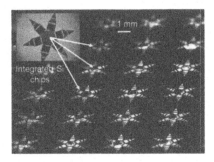
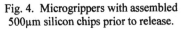

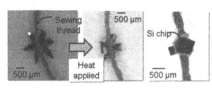

Fig. 4. Microgrippers with assembled 500μm silicon chips prior to release.

Fig. 5. Microgripper with assembled silicon chip, in water, induced to close around a piece of thread at 55°C.

parts quickly sank to the bottom of the ProTEK Remover 200 container, allowing us to dilute the solvent with IPA by a factor of 10^3, followed by dilution with DI water by another factor of 10^3.

For self-assembly, we suspended approximately 10^3 parts in approximately 3 ml ethylene glycol with 100 mM HCl to insure that the solder remained free of surface oxides and able to properly wet intended binding sites on adjacent parts. We placed heated the vial to 85°C, and began applying fluidic flow pulses of approximately 1 ml discharged through a Pasteur pipette every 1 s to stir the parts. We continued this agitation for 2 min. Whenever a part contacted a liquefied solder binding site, guided by the local magnetic force field immediately beneath that solder site, capillary forces from the solder held the part in place.

RESULTS AND DISCUSSION

Silicon Chips With Hinge Actuators

Figure 4 shows a result of assembling silicon chips with microgripper devices prior to release. The yield of this self-assembly step averaged 67%, and was as high as 91% shown in Fig. 4. Temperature and solubility constraints of the polymer hinges prevented the higher yields reported in other studies, but we are confident that further development will improve the yield.

Figure 5 demonstrates thermal actuation of a released microgripper in water as the temperature increased from 25° to 55°C. Although the polymer hinge layer normally transitions just above 40°C (an ideal range for in vivo applications), the release procedure likely altered the properties of the polymer, and higher temperatures were required to actuate the hinges.

Self-Assembly of 3D Circuit Elements

Prior to carrying out the fabrication process outlined in Fig. 3, we measured the magnetic hysteresis of the electroplated CoNiMnP material as shown in Fig. 6. This material exhibited an in-plane maximum BH product of 14.4 kJ/m^3. Figure 7 shows parts following most fabrication steps but prior to the final release steps, highlighting both the microscale CoNiMnP magnetic structures, and the successful coating of these structures with solder.

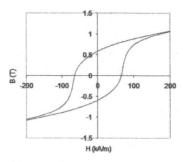

Fig. 6. Vibrating sample magnetometer B-H hysteresis curve for the CoNiMnP material electroplated and used to provide magnetic forces for self-assembling parts.

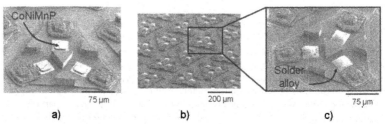

a)   b)   c)

Fig. 7. Scanning electron microscope images showing fabricated parts for the assembly shown in Fig. 1, prior to release. a) Image showing regions of CoNiMnP material electroplated and capped with a patterned Ti-Pt-Au solder wetting layer. b) Array of parts following the application of solder by dip coating. c) close up view of the dip-coated solder alloy.

Following the successful release of these parts from their carrier wafer, Fig. 8 shows initial results of the self-assembly procedure outlined in the experimental methods section. We defined self-assembly yield as the number of correct binding events relative to the total number of possible binding events. Each part had three possible binding sites on each side, for a total of six. Therefore two parts attached correctly at one side counted as two binding events (one for each part) out of 12 available binding sites. A group of three parts counted as four out of 18. Using this methodology on the groups of two and three correctly assembled parts in each image of Fig. 8 and several other captured images not shown here, the overall yield was 30/384 = 7.8%. Although rather low, it is a marked improvement over the less than 0.1% yield achieved with similar parts and capillary forces alone [30].

Although the parts were designed to physically restrict all unintended binding orientations, other orientations were possible and observed. Most wrong orientations were caused by unintended magnetic interactions. The parts used in the present experiment were actually etched to a much greater depth than expected during the first etch step, resulting in a very thin base of each part, evident in the images of Fig. 7. This thin layer caused magnets located at the undersides of each via and solder bump to lie at nearly the same plane, meaning that all magnetic interactions depicted in Fig. 1b) were of the same magnitude. Therefore, the bottom outer via of

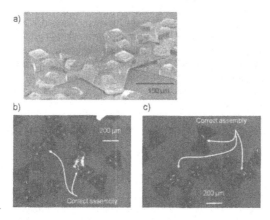

Fig. 8. Initial results for self-assembling triangular parts. a) Scanning electron microscope image showing a group of three properly-assembled parts. b) and c) Two microscope images showing successful assembly of several groups of two and three parts.

one part could easily attract the top outer via of another part, preventing the more permanent capillary bonds from ever forming, as two of the four apparently joined parts have done on the far left of Fig. 8B. Also, one part could simply "stack" on top of another part, as shown in the lower left of Fig. 8C by the group of three parts which actually contain a fourth part stacked in the middle. Other wrong orientations were caused by apparent north-north and south-south pole magnetic interactions, which may have been due to depolarization or other capillary effects overpowering the weaker magnetic interactions. In any case, we expect better control over fabrication tolerances to greatly reduce the rate of unintended binding, and lead to much better yield results.

We also expect to improve yield by trying different solder alloys and acid concentrations. For example, the alloy used in [30] had a much higher melting point. Consequently, the experiment was performed at a higher temperature, where lower acid concentrations could have had the same effects on cleaning solder surfaces of oxides. These oxides hinder the formation of capillary bonds, but the use of too much acid leads to excessive bubble formation at solder surfaces which also hinder the formation of these bonds. The latter phenomenon may have been partially responsible for the lower-than-expected yield of the present experiment.

CONCLUSIONS

In summary, we have demonstrated two examples of microscale systems using self-assembly driven by capillary forces, and magnetic forces in the case of the second example. In the first, thinned and diced 500 μm silicon chips were attached to thermally actuating hinges arranged to form microgrippers. The use of silicon illustrates a path towards integrating different types of silicon-based devices, and in fact we are targeting a specific, commercially-available microtransponder which is of the same size and form factor as the silicon chips used. In the second example, magnetic forces combined with capillary forces provided a route towards three dimensional circuit assembly. We have detailed the fabrication of fully-released microfabricated

159

parts designed to assemble in a particular fashion, including the incorporation of both permanently magnetic structures and a low melting point solder alloy. A yield of 7.8% was demonstrated, which is a significant improvement over past yields of less than 0.1% when capillary forces were used alone. Additional improvements are expected with fabrication and assembly process refinements.

ACKNOWLEDGEMENTS

K. E. Laflin acknowledges support from a Northrop Grumman Graduate Fellowship.

REFERENCES

1. C. J. Morris, S. A. Stauth, and B. A. Parviz, "Self-assembly for micro and nano scale packaging: steps toward self-packaging," *IEEE Trans. Adv. Packag.* **28**, 600–611, (2005).
2. S. Arscott, E. Peytavit, D. Vu, A. C. H. Rowe, and D. Paget, "Fluidic assembly of hybrid MEMS: a GaAs-based microcantilever spin injector," *J. Micromech. Microeng.* **20**, 025023, (2010).
3. R. J. Knuesel and H. O. Jacobs, "Self-assembly of microscopic chiplets at a liquid-liquid-solid interface forming a flexible segmented monocrystalline solar cell," *Proc. Natl. Acad. Sci.* **107**, 993–998, (2010).
4. R. J. Knuesel and H. Jacobs, "Fluidic surface-tension-directed self-assembly of miniaturized semiconductor dies across length scales and 3D topologies," in *MRS Spring Meeting, Symposium BB*, San Francisco, CA: Materials Research Society, April 13-17 2009.
5. X. Xiong, Y. Hanein, J. Fang, Y. Wang, W. Wang, D. T. Schwartz, and K. F. Böhringer, "Controlled multibatch self-assembly of microdevices," *J. Microelectromech. Sys.* **12**, 117–127, (2003).
6. S. A. Stauth and B. A. Parviz, "Self-assembled single-crystal silicon circuits on plastic," *Proc. Natl. Acad. Sci. U. S. A.* **103**, 13922–13927, (2006).
7. U. Srinivasan, D. Liepmann, and R. T. Howe, "Microstructure to substrate self-assembly using capillary forces," *J. Microelectromech. Sys.* **10**, 17–24, (2001).
8. U. Srinivasan, M. Helmbrecht, C. Rembe, R. Muller, and R. Howe, "Fluidic self-assembly of micromirrors onto microactuators using capillary forces," *IEEE J. Sel. Top. Quantum Electron.* **8**, 4–11, (2002).
9. H. O. Jacobs, A. R. Tao, A. Schwartz, D. H. Gracias, and G. M. Whitesides, "Fabrication of a cylindrical display by patterned assembly." *Science* **296**, 323–5, (2002).
10. J.-H. Cho and D. H. Gracias, "Self-assembly of lithographically patterned nanoparticles," *Nano Lett.*," (2009).
11. T. G. Leong, P. A. Lester, T. L. Koh, E. K. Call, and D. H. Gracias, "Surface tension-driven self-folding polyhedra," *Langmuir* **23**, 8747–8751, (2007).
12. W. Zheng, P. Buhlmann, and H. O. Jacobs, "Sequential shape-and-solder-directed self-assembly of functional microsystems," *Proc. Natl. Acad. Sci. U. S. A.* **101**, 12814–12817, (2004).
13. W. Zheng and H. O. Jacobs, "Shape-and-solder-directed self-assembly to package semiconductor device segments," *Appl. Phys. Lett.* **85**, 3635–3637, (2004).
14. T. D. Clark, J. Tien, D. C. Duffy, K. E. Paul, and G. M. Whitesides, "Self-assembly of 10-μm-sized objects into ordered three-dimensional arrays," *J. Am. Chem. Soc.* **123**, 7677–7682, (2001).

15. T. D. Clark, R. Ferrigno, J. Tien, K. E. Paul, and G. M. Whitesides, "Template-directed self-assembly of 10-μm-sized hexagonal plates," *J. Am. Chem. Soc.* **124**, 5419–5426, (2002).
16. C. G. Fonstad, Jr. and M. Zahn, "Method and system for magnetically assisted statistical assembly of wafers," United States Patent: 6,888,178, May 2005.
17. R. Rivero, S. Shet, M. Booty, A. Fiory, and N. Ravindra, "Modeling of magnetic-field-assisted assembly of semiconductor devices," *J. Electron. Mater.* **37**, 374–378, (2008).
18. Q. Ramadan, Y. S. Uk, and K. Vaidyanathan, "Large scale microcomponents assembly using an external magnetic array," *Appl. Phys. Lett.* **90**, 172502–3, (2007).
19. H. Ye, Z. Gu, T. Yu, and D. H. Gracias, "Integrating nanowires with substrates using directed assembly and nanoscale soldering," *IEEE Trans. Nanotechnol.* **5**, 62 – 6, (2006).
20. J. C. Love, A. R. Urbach, M. G. Prentiss, and G. M. Whitesides, "Three-dimensional self-assembly of metallic rods with submicron diameters using magnetic interactions." *J. Am. Chem. Soc.* **125**, 12696–7, (2003).
21. T. G. Leong, C. L. Randall, B. R. Benson, A. M. Zarafshar, and D. H. Gracias, "Self-loading lithographically structured microcontainers: 3D patterned, mobile microwells," *Lab Chip* **8**, 1621–1624, (2008).
22. N. Bassik, G. M. Stern, and D. H. Gracias, "Microassembly based on hands free origami with bidirectional curvature," *Appl. Phys. Lett.* **95**, 091901–3, (2009).
23. T. G. Leong, C. L. Randall, B. R. Benson, N. Bassik, G. M. Stern, and D. H. Gracias, "Tetherless thermobiochemically actuated microgrippers," *Proc. Natl. Acad. Sci.* **106**, 703–708, (2009).
24. J. S. Randhawa, T. G. Leong, N. Bassik, B. R. Benson, M. T. Jochmans, and D. H. Gracias, "Pick-and-place using chemically actuated microgrippers," *J. Am. Chem. Soc.* **130**, 17238–17239, (2008).
25. S. Guan and B. J. Nelson, "Electrodeposition of low residual stress conimnp hard magnetic thin films for magnetic mems actuators," *J. Magn. Magn. Mater.* **292**, 49 – 58, (2005).
26. H. J. Cho and C. H. Ahn, "A bidirectional magnetic microactuator using electroplated permanent magnet arrays," *J. Microelectromech. Sys.* **11**, 78–84, (2002).
27. T. M. Liakopoulos, W. Zhang, and C. H. Ahn, "Micromachined thick permanent magnet arrays on silicon wafers," *IEEE Trans. Magn.* **32**, 5154–5156, (1996).
28. J. Horkans, D. J. Seagle, and I. C. H. Chang, "Electroplated magnetic media with vertical anisotropy," *J. Electrochem. Soc.* **137**, 2056–2061, (1990).
29. M. Grapes and C. J. Morris, "Optimizing the CoNiMnP electrodeposition process using taguchi design of experiments," *J. Electrochem. Soc.,* (2010), submitted.
30. C. J. Morris and M. Dubey, "Toward three dimensional circuits formed by molten-alloy driven self-assembly," in *Proc. 26th Annual Army Science Conference*, Orlando, FL, Dec. 1-4 2008.

Mater. Res. Soc. Symp. Proc. Vol. 1272 © 2010 Materials Research Society   1272-OO06-05-LL06-05

A moth-eye bio-inspired approach to planar isotropic diffraction

Petros I Stavroulakis, Stuart A Boden and Darren M Bagnall[1]
[1]Nano Research Group, Electronics and Computer Science, University of Southampton, SO17 1BJ, United Kingdom

ABSTRACT

A regular hexagonally packed biomimetic moth-eye antireflective surface acts as a diffraction grating at short wavelengths of the visible spectrum and shallow angles of incidence. These gratings display strong backscattered iridescence with 6-fold optical symmetry. The optical symmetry of real moth eyes is effectively infinite as nature utilizes large number of uniquely oriented domains. In this work we report on a biomimetic moth-eye surface created via nanosphere lithography with a very large distribution of close-packed tessellated domains and the resulting optical symmetry is compared to that of another widely known highly isotropic diffraction grating, also inspired by nature, the sunflower pattern.

A white-light laser reflectometry system is used to measure and compare the diffraction pattern isotropy from both structures. The tessellated close-packed structure diffraction pattern approaches that of infinite optical symmetry even though the underlying pattern only possesses a six-fold symmetry. Hence, the angular isotropy observed for the sunflower pattern is replicated to a large extent via a self-assembly procedure, whilst circumventing the complicated design and manufacturing requirements of the sunflower pattern.

INTRODUCTION

The cornea of some species of butterfly and nocturnal moth (Figure 1) are covered in subwavelength protuberances which produce an effective grated refractive index antireflection layer [1]. This layer is known as a 'moth-eye' antireflection layer, first noted by Bernard [2] on the cornea of a moth's eye.

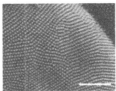

Figure 1 SEM Image of the corneal nipple arrays in the peacock butterfly (Inachis Io). The scale bar is 2μm. Source [3]

It is possible to use nanosphere lithography to create periodic close-packed (hexagonal or trigonal packing) structures which are very similar to the small domains that exist on the cornea of the eyes of many Lepidoptera [3]. Self-assembly methods are much more time and cost effective than conventional nanofabrication techniques for rapidly creating large (>100μm^2) nanoengineered surfaces [4]. Hence, developing a method of achieving an equivalent isotropic behavior to the sunflower pattern but with a cheap and parallel self-assembly technique would be very useful for use in commercial optical applications that require high optical symmetry such as low-index contrast photonic crystals [5] or extracting more light from LED's [6].

The sunflower pattern is another biomimetic pattern has been inspired by the positioning of the florets on a sunflower's head [7]. It is one of the most optically isotropic two-dimensional patterns available, displaying a complete ring in its 2D Fourier transform [8]. However, the complicated sunflower pattern cannot be readily implemented via self-assembly methods because the relative locations of the features required to create the pattern are too complicated for any artificial self-assembly procedure known to date (other than those used by nature).

The biomimetic pattern shown in Figure 1 is referred to here as a 'tessellated close-packed pattern'. Single-orientation, close-packed structures have been previously studied for antireflection applications [9], as have been self-assembled tessellated close-packed surfaces [10][11] such as those that will be shown in this work. The optical characterization of these surfaces however, has mostly been limited to normal incidence reflectance measurements of the samples whereas the effect of the pattern tessellation on the diffraction orders that emerge from the lattice at high angles of incidence is usually ignored.

In this work, emphasis will be put on the characterization of the diffraction properties of such layers, and not on their antireflection properties. It will be shown that the Fourier transform of these self-assembled structures exhibits a near-isotropic behavior depending on the distribution of the alligment of the six-fold symmetric domains [8].

EXPERIMENTAL DETAILS

The nanosphere monolayer mask was self-assembled via a liquid surface assembly procedure that works by depositing a nanosphere colloid on the surface of ultrapure deionized water stored in a Teflon container. Using this apparatus, carboxylate-modified 780nm–diameter sized polystyerene nanospheres (Duke Scientific) were deposited in tessellated close-pack patterns onto 2x2cm silicon sample areas. The nanosphere deposition technique is based on a method described previously [4] was optimized for our own work [12] so that a monolayer of nanospheres in tessellated close packed structures can be assembled in large areas on silicon substrate >1cm^2.

The nanospheres were then etched in the radial direction using oxygen plasma etch recipe in the Reactive Ion Etcher (Oxford Instruments, Plasmalab80) whilst maintaining their initial positions. This allows for reduction of the diameter of the spheres without changing the pattern period. This step is essential to separate the spheres from each other in order to allow anisotropic etching of the substrate into a forest of pillar features. The subsequent step of anisotropic etching of the substrate was realized via DRIE, (Deep Reactive Ion Etching) as shown in [13]. This anisotropic etch procedure is selective to the substrate and etches away the exposed silicon material creating silicon rods with a diameter equal to that of the spheres in the nanosphere mask. The spheres are then dissolved in a subsequent long oxygen etch via O2 plasma etching (Figure 2a).

The sunflower and close-packed patterns were defined by e-beam lithography (Figure 2 b, c) in PMMA. The samples then underwent a subsequent mask developing and anisotropic etching procedure to create a cylindrical pillar profile of pattern features, similar to that created by nanosphere lithography.

Figure 2 SEM Images of (a) the profile of the pillars created by nanosphere self assembly and DRIE etching (b) sunflower pattern and (c) close-packed pattern both created.

When comparing the Fourier transforms of different magnifications of the nanosphere-manufactured self-assembled patterns we notice that as magnification increases from 1000x to 6500x, the Fourier transform of the pattern changes from resembling that of a very isotropic ring to one having pronounced six-fold symmetry (Figure 3). Hence, the diffraction pattern of this self-assembled biomimetic pattern is expected to be quite isotropic when an area >80µm^2 is probed. A custom reflectometry apparatus was used for the experiment. The white light laser source that was available (Fianium, UK) has a much larger beam diameter (~1mm) at normal incidence and so it was able to pick up the isotropic diffraction behavior of the sample easily. The spectrum acquisition was done through a fibre optic cable attached to a detector arm, which led to a spectrometer (B&W Tek). Thus, it is verified that the self-assembled tessellated close packed pattern created possesses a very high rotational symmetry which can be verified with the laser apparatus available.

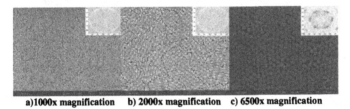

a)1000x magnification   b) 2000x magnification   c) 6500x magnification

Figure 3. SEM images (top view) of different magnifications of nanosphere self-assembled diffraction grating patterns on silicon with respective Fourier transforms on top right corner.

The patterns that were used in this comparative analysis have diffraction grating periods of, 350nm for the monocrystalline close packed pattern, 780nm for the tessellated close packed sample and 560nm for the sunflower sample. These values were extracted from the design files that were used to manufacture them. The values were verified by inferring the grating period from an angular scan of the diffraction spectrum for all three samples and fitting the diffraction equation to the experimental results. The range of angles of incidence that can be used to create diffraction of the -1 order in the visible spectrum (400-700nm) for samples with these periods can be found from the equation of diffraction maxima [14]. It was found that diffraction in the visible spectrum is possible for the close-packed sample which has the smallest period, at angles of incidence around 60 degrees whereas the sunflower and tessellated close-packed samples have a large enough period to diffract the -1 order at normal incidence.

165

Hence, the nanosphere and sunflower samples were placed on the setup and rotated in azimuth at normal incidence and the close packed sample was rotated at 60 degrees angle of incidence. The detector was placed accordingly to monitor how the intensity of the diffracted wavelength changes as the sample is rotated in the azimuth and the results were recorded (Figure 4). For the nanosphere and sunflower samples the monitored diffraction wavelength was 423nm whereas for the close-packed sample the monitored diffraction wavelength was 453nm.

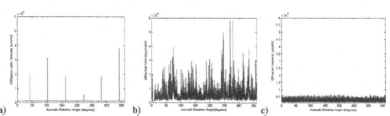

a)   b)   c)

Figure 4. Power variation vs azimuth angle of rotation for (a) close-packed, (b) nanosphere and (c) sunflower samples.

DISCUSSION

The diffractive properties of biomimetic structures that possess order which is confined to their nearest neighbors correlates with their Fourier Transform, as also is the case with patterns that retain order throughout the whole surface they cover [15]. A unique property of tessellated close-packed surface is that their diffractive symmetry is tunable and proportional to the amount of orientations considered in the pattern. This means that the rotational symmetry of the pattern increases linearly, the more orientations of the base pattern are included. As Figure 5 depicts, by increasing the number of orientations of the close-packed pattern in the structure from 1-4, the number of peaks noticed in the Fourier Transform, increase from 6-24.

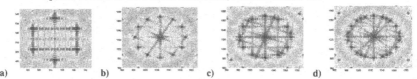

a)   b)   c)   d)

Figure 5. Fourier transforms of simulation patterns comprising of one(a), two(b), three(c), four(d) orientations of close-packed patterns. Axes display the pixel numbers in the FFT image.

It was found that in order for an orientation to contribute in increasing the overall optical diffraction symmetry of the pattern, the angular offset between all the pattern orientations and the reference orientation has to be unique and smaller than the angle of repeating symmetry of the underlying pattern which in this case is six-fold symmetric and hence the repeating angle is 60 degrees (Figure 6a). It can be projected therefore that the diffraction pattern will keep increasing linearly in symmetry the more such unique orientations of the base pattern we take into account. As previously noted, large area tessellated close packed structures with patterns that are oriented in lots of different orientations were produced in large areas (~2cm^2) via nanosphere lithography at room temperature. The angular offset and area distribution of the orientations of the close-packed pattern in the tessellated domains cannot be fully controlled because they depend on the specific molecular dynamics of each self-assembly run. However, by taking Fourier Transforms of the SEM images from the constructed sample it is possible to infer the angular isotropy of the diffraction pattern (Figure 6c). Indeed, from the Fourier transforms of the SEM images that were taken for the nanosphere sample it is shown that a very isotropic pattern was achieved. The underlying non-tessellated close pack pattern in contrast has a strong six-fold symmetry (Figure 6b) as depicted in its 2D Fourier Transform, whereas the sunflower pattern in comparison also depicts a complete ring (Figure 6d) corresponding to a very isotropic diffraction pattern.

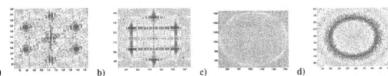

a)   b)   c)   d)

Figure 6. Fourier transforms of a) the simulated tessellated pattern with 2 close-packed orientations offset 60 degrees to each other and Fourier transforms of the SEM Images of the b) close-packed, c) tessellated close-packed and d) sunflower samples that were manufactured. Axes display the pixel numbers in the FFT image.

As it is shown in the azimuth scan of the -1 diffraction order (Figure 4a) for the close-packed pattern, the number of diffraction peaks are six, as was expected from the Fourier transform. The sample created via nanosphere lithography on the other hand has countless diffraction peaks, even though the underlying pattern in the tessellated domains is the same (Figure 4b). In comparison, the sunflower pattern is also isotropic but more homogeneous with respect to the power produced at each azimuth diffraction angle, with countless peaks at similar power levels in the azimuth plane (Figure 4c). The polar plots in (Figure 7a, b) depict the difference between the optical symmetry of these patterns in a qualitative comparison ignoring diffraction power. It is shown that the isotropic nature of the nanosphere sample is very similar to the sunflower sample (Figure 7c) and does not at all resemble that of the close packed structure.

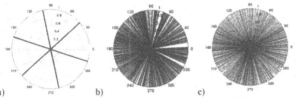

a)   b)   c)

Figure 7. Polar plots comparing hexagonal (a), nanosphere (b) and sunflower (c) azimuth diffraction angles

CONCLUSIONS

Tessellated close-packed samples which are normally undesired in nanosphere lithography applications, exhibit optical symmetry which can be controllably varied. It has been verified via Fourier transform analysis and optical experiment that tessellated close-packed moth-eye inspired diffraction gratings created by nanosphere lithography possess a large number of orientations of the close-packed pattern which in turn equates to a very high optical symmetry which is comparable to that of the isotropic sunflower pattern. In order to achieve complete 2D isotropy using this new biomimetic concept, homogeneous distribution of the various orientations of the close-packed domains is required, something which could not be achieved the method of manufacture shown here. Investigation in achieving complete isotropy via nanosphere lithography will be pursued in future work.

REFERENCES

[1]   S.J. Wilson and M.C. Hutley, "The Optical Properties of 'Moth Eye' Antireflection Surfaces," *Journal of Modern Optics*, vol. 29, 1982, pp. 993-1009.

[2]   C.G. Bernard, "Structural and functional adaptation in a visual system," *Endeavour*, vol. 26, 1967, pp. 79-84.

[3]   D.G. Stavenga, S. Foletti, G. Palasantzas, and K. Arikawa, "Light on the moth-eye corneal nipple array of butterflies.," *Proceedings. Biological sciences / The Royal Society*, vol. 273, 2006, pp. 661-7.

[4]   S.M. Weekes, F.Y. Ogrin, W.A. Murray, and P.S. Keatley, "Macroscopic Arrays of Magnetic Nanostructures from Self-Assembled Nanosphere Templates," *Langmuir*, vol. 23, 2007, pp. 1057-1060.

[5]   M.E. Pollard and G.J. Parker, "Low-contrast bandgaps of a planar parabolic spiral lattice," *Optics Letters*, vol. 34, 2009, p. 2805.

[6]   S. Fan, P. Villeneuve, J. Joannopoulos, and E. Schubert, "High Extraction Efficiency of Spontaneous Emission from Slabs of Photonic Crystals," *Physical Review Letters*, vol. 78, 1997, pp. 3294-3297.

[7]   H. Vogel, "A better way to construct the sunflower head," *Mathem. Biosci.*, vol. 44, 1979, pp. 179-189.

[8]   M. Mihailescu, A. Preda, D. Cojoc, E. Scarlat, and L. Preda, "Diffraction patterns from a phyllotaxis-type arrangement," *Optics and Lasers in Engineering*, vol. 46, 2008, pp. 802-809.

[9]   S.A. Boden and D.M. Bagnall, "Tunable reflection minima of nanostructured antireflective surfaces," *Applied Physics Letters*, vol. 93, 2008, p. 133108.

[10]   H. Xu, N. Lu, D. Qi, L. Gao, J. Hao, Y. Wang, and L. Chi, "Broadband antireflective Si nanopillar arrays produced by nanosphere lithography," *Microelectronic Engineering*, vol. 86, 2009, pp. 850-852.

[11]   Q. Chen, G. Hubbard, P.A. Shields, C. Liu, D.W. Allsopp, W.N. Wang, and S. Abbott, "Broadband moth-eye antireflection coatings fabricated by low-cost nanoimprinting," *Applied Physics Letters*, vol. 94, 2009, p. 263118.

[12]   P.I. Stavroulakis, N. Christou, and D. Bagnall, "Improved deposition of large scale ordered nanosphere monolayers via liquid surface self-assembly," *Materials Science and Engineering: B*, vol. 165, 2009, p. 186–189.

[13]   C.L. Cheung, R.J. Nikolić, C.E. Reinhardt, and T.F. Wang, "Fabrication of nanopillars by nanosphere lithography," *Nanotechnology*, vol. 17, 2006, pp. 1339-1343.

[14]   M. Born and E. Wolf, *Principles of Optics*, Page 449, Cambridge University Press, 2002.

[15]   R.O. Prum and R.H. Torres, "A Fourier tool for the Analysis of Coherent Light Scattering by Bio-Optical Nanostructures," *Integr. Comp. Biol.*, vol. 43, 2003, pp. 591-602.

Interfacing Biomolecules and Functional (Nano) Materials

Mater. Res. Soc. Symp. Proc. Vol. 1272 © 2010 Materials Research Society
1272-PP03-06

Growth process and crystallographic properties of ammonia-induced vaterite

Qiaona Hu[1], Jiaming Zhang[1], and Udo Becker[1]
[1]Department of Geological Sciences, University of Michigan, Ann Arbor, MI 48109, USA

ABSTRACT

Metastable vaterite crystals were produced by increasing the pH of Ca^{2+}- and CO_3^{2+}-containing solutions through diffusing ammonia gas. The SEM and TEM studies indicate that this ammonia-induced vaterite is polycrystalline with a 6-fold symmetry of the crystal aggregate. The morphology and crystallographic properties of this assemblage change during crystallization. One hour after nucleation starts, vaterite grains display a spherical structure composed of nano-particles (5-10 nm) with random crystallographic orientations. After that, horizontal layers begin to develop at the edge of the sphere and gradually tilt toward the center as they grow vertically, which results in a three-dimensional morphology with a dent in the center. The vaterite grains mature fully 16 hours after nucleation. TEM analysis indicates the grown vaterite grain (50-60 μm) consists of numerous hexagonal pieces of single crystals (1-2 μm) of similar crystallographic orientations. High-resolution TEM demonstrates that these single crystals grow along (001) with {110} hexagonal boundaries.

INTRODUCTION

Calcium carbonate ($CaCO_3$) occurs widely on the Earth's surface and represents the largest geochemical reservoir for carbon.[1, 2] Vaterite, a less thermodynamically stable polymorph of $CaCO_3$, has been of great interest because its crystallization is strongly associated with biogenic activities and itself an important precursor in several carbonate-forming processes. Due to its large surface area and high surface energy, vaterite is also largely applied as dispersant in industries, such as paper-making, oil-field drilling fluids, and regenerative medicine.[3, 4] Therefore, understanding the vaterite growth process is important to both the materials industry and to the research of biomineralization mechanisms.

Previous studies have intensively reported on vaterite formation induced by organic additives or precipitating it from salt solution that were highly supersaturated with respect to $CaCO_3$.[5-10] However, only few studies have focused on the ability of inorganic additives promoting vaterite growth.[11] The growth process and the crystallographic properties of vaterite crystals have also been poorly documented. In this project, the growth process of vaterite and the function of ammonia as an inorganic additive in vaterite formation have been investigated.

EXPERIMENTAL PROCEDURE

Vaterite was produced by increasing the pH of Ca^{2+} and-CO_3^{2-} (0.002 M) containing solutions by diffusing ammonia gas into solution and, thus, supersaturating the solution with respect to calcium carbonate at ambient conditions. The original pH was adjusted to 3.4 by adding 10% HCl. It was found that the polymorphic composition of the $CaCO_3$ changes with the diffusion rate of ammonia. In this study, the NH_3 diffusion rate was adjusted to increase the pH of the salt solution in one hour from the original value, 3.4, to the final one, 9.6. At this pH, the carbonate concentration is high enough to precipitate vaterite. Control run were also conducted to determine whether NH_3 promotes vaterite formation or if NH_3 only plays the role as pH modifier. All experimental settings were the same in the control experiment, except for applying dilute NaOH solution, instead of NH_3 gas, to increase the pH of the salt solution. The rate of pH increase by adding NaOH was the same as with the ammonia method in order to avoid potential kinetic differences which would influence the polymorph composition. It was found that the precipitation in the control run was almost pure calcite, indicating the ability of ammonia to promote vaterite precipitation.

X-ray diffraction (XRD) with subsequent Rietveld refinement was used to confirm that the precipitate was indeed vaterite. The structural and other properties were further characterized using SEM and field-emission-gun SEM. The development of the vaterite morphology was observed using optical microscopy *in situ* every two hours throughout the mineralization. To study the nucleation and growth process, as well as the corresponding crystallographic properties after the solution reached the final pH, vaterite grains were harvested and analyzed using transmission electron microscope every 30 minutes in the first four hours and every two hours thereafter until the crystals matured (after 16 hours).

RESULTS AND DISCUSSIONS

Morphology of vaterite at different growth stages

Figure 1 shows a series of scanning electron microscope (SEM) images displaying the typical vaterite morphology in different growth stages, illustrating a dramatic shape change as a function of time. The labels, marked on the right corner of each image, indicate how much time passed after the solution reached the final pH when the crystals were collected for SEM analysis. Vaterite grains exhibit a spherical shape and are about 2 μm in diameter after the first hour. After 1-1.5 hours, horizontal layers start to develop, and the grain turns into a hexagonal plate-like shape about 10 μm in diameter. After that, the crystal keeps developing and the layers at the periphery grow faster than the ones in central area, which causes the growing layers to tilt progressively toward the center, resulting in a three-dimensional (3D) morphology with a dent in the center. This grain reaches the size of about 20 μm in diameter after 4 hours, and about 60 μm after 16 hours when it gets fully matured. The grown vaterite displays a layered structure of 6-fold symmetry, with the latest layers almost vertically-embracing a void core (Figure 1E). These layers are shown with higher magnification in Figure 1F.

Comparing Figures 1A and 1F (both at the same scale), it is found that in addition to the morphology, the crystallography of vaterite forming in the first hour is different from the one of the matured vaterite. Instead of a crystalline structure, which is found in a mature grain, the initial grain appears like an aggregation of poorly-orientated nanoparticles. In order to analyze the detailed crystallographic properties of vaterite in the early and final crystallization stages (first hour vs. 16 hours), TEM characterization was applied.

172

Crystallographic information of vaterite grain in the early and final stages

For the spherical vaterite grain forming in the first hour, bright and dark field images taken at the same area illustrate that the grain is composed of nanoparticles with different crystallographic orientations (Figure 2A and B). The corresponding high-resolution TEM image confirms the discontinuous lattices of these nano-crystals (5 nm) with different color contrasts. The fast Fourier transformation pattern of the high-resolution TEM image shows the crystal structure of the grain (Figure 2C) including the internal structure and the superimposed polycrystalline pattern of vaterite.

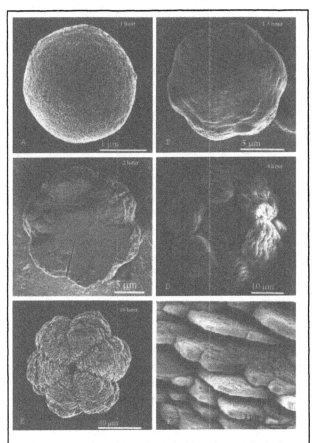

Figure 1. SEM images displaying the growth procedure of vaterite grain as a function of time

173

Another interesting fact is that vaterite of this nano-crystalline structure possesses high resistance to the ionizing irradiation of the electronic beam under the TEM characterization. In general, $CaCO_3$ is vulnerable to high temperature and electron bombardment. Only after irradiation at a fluence of $1.5 - 2.5 \cdot 10^2$ electrons/cm^2, both single crystals of calcite (20-30 μm) or crystalline layers of grown vaterite (1-2 μm) begin to decompose. However, the nanocrystalline vaterite kept stable even after an irradiation at a fluence of $9 \cdot 10^3$ electrons/cm^2. The high-resolution TEM image (Figure 2D) displays the lattice of nanocrystal vaterite which is almost the same as the original one before the $9 \cdot 10^3$ electrons/cm^2 irradiation (Figure 2C).

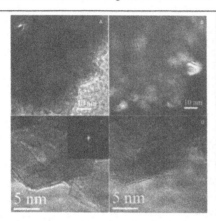

Figure 2. A and B: bright and dark field images, respectively, showing the vaterite nanocrystalline structure lack of uniform orientations forming at the first hour after the solution pH reaching the maximum value. C: corresponding high resolution TEM image. D: corresponding high resolution TEM image after the 10-minute electron beam bombardment, displaying high resistance to the ionizing irradiation of the electron beam

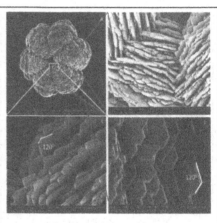

Figure 3. A and B: Vaterite grain is composed of larger flakes (10-20 μm). C&D: flakes are not continuously growing layers but built by numerous smaller hexagonal pieces, each about 1 μm in diameter.

In contrast, with vaterite forming in the early stage, the matured vaterite grain is built by single crystals of hexagonal shape (1 μm) with more or less parallel orientations. Low-magnification SEM shows that each vaterite grain is composed of larger flakes, each about 10-20 μm in diameter (Figure 3 B). High-magnification SEM and TEM images reveal that these flakes are not continuously growing layers but built by numerous smaller hexagonal pieces of ~1 μm in diameter (Figure 3C and D, Figure 4D). The corresponding diffraction patterns of the high-magnification TEM image (taken from the red square area) demonstrates that each hexagonal piece is a single crystal of vaterite with (001) basal surfaces and hexagonal boundaries terminated by {110} steps. A periodic arrangement of atoms of the hexagonal crystallites is

174

revealed by the high-resolution TEM image taken at the center area of the hexagonal platelet, which shows the long-range order of atoms on the vaterite (001) surface (Figure 4C).

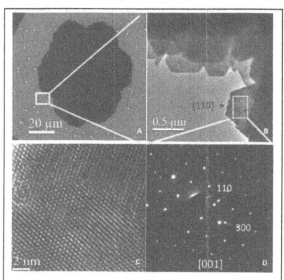

Figure 4.   A and B: low and high magnification TEM images show the layered structure of vaterite and the unit bricks that are small hexagonal pieces of 1 μm. C: a high resolution TEM image shows atomic lattices of vaterite along [001], revealing an ordered arrangement of atoms. D: corresponding diffraction pattern along vaterite [001], demonstrating the hexagonal pieces shown in B terminates at {110}.

It is interesting to note that, although the strong diffraction spots close to the transmitted spot (in the center of the TEM diffraction pattern) are overlapping, starting from the third nearest circle (*i.e.* (400) spots), two separated dots can be detected at each location of the diffraction spot.   The distances between the two dots increase with distance from the center (as illustrated). The contrasted color of the high-resolution TEM image (Figure 4B) demonstrates that the diffraction patterns of two small platelets are superimposed at the location where the diffraction pattern was taken (as illustrated).   Therefore, the Figure 4D actually depicts the crystallographic structures of two crystals, illustrating two phenomena: 1) these two crystallites have the same structure, since the two sets of patterns are identical, and 2) the two crystallographic orientations diverge a bit from each other because one diffraction pattern is rotated slightly with respect to the other.

The hexagonal shape of the grown vaterite seems to be a twin but actually is not. The magnification of the contact area of adjoining "petals" reveals discontinuous connections (Figure 3B), and the extinction angles of the adjacent "petals" vary from grain to grain under the crossed polarized light. Both phenomena disagree with the characteristics of twinned crystals.

CONCLUSIONS

This paper reveals a change of morphology and crystallography of vaterite during crystallization and the promotion of vaterite growth in the presence of ammonia. The vaterite grain forming in the first hour is a sphere (2 μm) composed of nano-particle random orientations, while the grown vaterite grain is a layered structure with 6-fold symmetry (50 μm) assembled by single crystals of similar orientations. The nano-crystalline vaterite grains forming in the early stage have high resistance to electron bombardment in the TEM.

REFERENCES

1. O. Braissant, G. Cailleau, C. Dupraz and E. P. Verrecchia, Journal of Sedimentary Research **73** (3), 485-490 (2003).
2. S. E. Wolf, N. Loges, B. Mathiasch, M. Panthofer, I. Mey, A. Janshoff and W. Tremel, Angewandte Chemie International Edition **46** (29), 5618-5623 (2007).
3. G. Hadiko, Y. S. Han, M. Fuji and M. Takahashi, presented at the Asian International Conference on Advanced Materials, Beijing, Peoples R China, 2005 (unpublished).
4. K. Fuchigami, Y. Taguchi and M. Tanaka, Advanced Powder Technology **20** (1), 74-79 (2009).
5. T. Kasuga, H. Maeda, K. Kato, M. Nogami, K.-i. Hata and M. Ueda, Biomaterials **24** (19), 3247-3253 (2003).
6. M. Kitamura, Journal of Colloid and Interface Science **236** (2), 318-327 (2001).
7. T. Ogino, T. Suzuki and K. Sawada, Geochimica Et Cosmochimica Acta **51** (10), 2757-2767 (1987).
8. D. Kralj, L. Brecevic and J. Kontrec, J. Cryst. Growth **177** (3-4), 248-257 (1997).
9. H. Colfen and M. Antonietti, Langmuir **14** (3), 582-589 (1998).
10. S. E. Grasby, Geochimica Et Cosmochimica Acta **67** (9), 1659-1666 (2003).
11. N. Gehrke, H. Colfen, N. Pinna, M. Antonietti and N. Nassif, Crystal Growth & Design **5** (4), 1317-1319 (2005).

Mater. Res. Soc. Symp. Proc. Vol. 1272 © 2010 Materials Research Society 1272-PP03-11

Lipase Immobilized within Novel Silica-based Hybrid Foams: Synthesis, Characterizations and Catalytic Properties

Nicolas Brun[1,2], Annick Babeau Garcia[1], Victor Oestreicher[1], Hervé Deleuze[2], Clément Sanchez[3] and Rénal Backov[1]

[1]Centre de Recherche Paul Pascal, CNRS, Pessac, France.
[2]Institut des Sciences Moléculaires, CNRS – Université de Bordeaux, Talence, France.
[3]Laboratoire de Chimie de la Matière Condensée de Paris, CNRS – Université Pierre et Marie Curie, Paris, France.

ABSTRACT

The covalent immobilization of crude lipases within silica-based macroporous frameworks have been performed by combining sol-gel process, concentrated direct emulsion, lyotropic mesophase and post-synthesis functionalizations. The as-synthesized open cell hybrid monoliths exhibit high macroscopic porosity, around 90 %, providing interconnected scaffold while reducing the diffusion low kinetic issue. The entrapment of enzymes in such foams deals with a high stability over esterification and transesterification batch process catalysis.

INTRODUCTION

Immobilization or entrapment of biocatalysts [1] onto or within porous materials by either physical adsorption [2], covalent attachment [3], inclusion or encapsulation by sol-gel route [4], represent an attractive and efficient approach to facilitate their use in continuous processes. Such systems are expected to enhance stability, activity and selectivity while allowing efficient separation, recycling and reuse of costly enzymes. Thus, design of new functional porous materials to immobilize active biomacromolecules is both of economic and ecologic interests. Recently, as emerged the novel concept of "integrative chemistry" [5], from the interface between bio-inspired approaches and hybrid organic-inorganic chemistry. Through the application of this concept, the assembling of a large variety of molecular precursors or nanobuilding blocks into engineered hierarchical structures should be strongly pre-dictated. Particularly, functional ordered macro-mesoporous materials are of interest for multiple applications in heterogeneous catalysis, separation techniques, purification of wastewaters, sensors, optics etc. With this aim, our research group has developed a way to obtain hybrid macrocellular silica-based monoliths, labeled "Organo-Si(HIPE)" (acronym refers to the High Internal Phase Emulsion process [6]), exhibiting a hierarchically structured porosity in view of reaching final polyfunctionalities [7]. With the same strategy, and using the concept of immobilized biocatalysts, we have just designed a new series of biohybrid foams labelled Lipase@Organo-Si(HIPE), bearing high catalytic performances [8].

EXPERIMENT

Synthesis. *Si(HIPE) synthesis.* Typically, tetraethyl orthosilicate (TEOS; 5 g) was added to an aqueous solution of tetradecyltrimethylammonium bromide (TTAB; 16 g, 35 wt %) previously acidified (7 g of HCl). Hydrolysis was left going on until a monophasic medium was obtained. The oily phase constituted of dodecane (35 g) was then emulsified drop by drop into the hydrophilic continuous phase using a mortar, and the emulsion was allowed to condense for 1 week at room temperature. The as-synthesized monoliths were washed three times with a tetrahydrofuran/acetone mixture (1:1 v/v) to extract the oily phase. Drying of the materials for a week at room temperature was followed by a thermal treatment at 650 °C (heating rate of 2 °C.min^{-1}) for 6 h, with a 2 h-plateau at 200 °C.

Grafted Glymo-Si(HIPE) synthesis. A piece of native Si-(HIPE), (1 g) was added to a solution of (3-Glycidyloxypropyl)trimethoxysilane organosilane (Glymo) (0.01 mol) in chloroform (120 g). The suspension was placed under vacuum for a good impregnation. After 48 hours aging at room temperature, the solution is filtrated, and then the monoliths washed with chloroform and acetone and dried in air.

Lipase immobilization. The enzymatic solutions were prepared as follow. Crude lipase from *Candida Rugosa* (540 mg; powder; 3%wt. of enzyme; E.C.3.1.1.3, Type VII, 700 U/mg, Sigma) was dispersed into 18 ml of distilled water, while crude lipase from *Thermomyces Lanuginosus* (53 mg; solution, 24.5 mg of enzyme / mL; ≥100,000 U/g, Sigma) is dissolved into 10 mL distilled water. The immobilizations of lipases were carried out by introducing a piece of Glymo-Si(HIPE) open cell monolith (around 300 mg) into the enzymatic solutions. A vacuum (20 mbar) was applied during 72 hours at ambient temperature. The immobilized lipases were then extracted from the impregnation media and washed three times with distilled water in order to remove the enzyme excess. Finals [C-*CRl*]@Glymo-Si(HIPE) and [C-*TLl*]@Glymo-Si(HIPE) heterogeneous catalysts are then dried at room temperature during 12 hours and stocked in a fridge at 4°C.

Characterization. Scanning Electron Microscopy (SEM) observations were performed with a Hitachi TM-1000 apparatus at 15 kV. Surface areas and pore characteristics were obtained with a Micromeritics ASAP 2010. Intrusion/extrusion mercury measurements were performed using a Micromeritics Autopore IV apparatus this to reach the scaffolds macrocellular cells characteristics. TEM experiments were performed with a Jeol 2000 FX microscope (accelerating voltage of 200 kV). The samples were prepared as follows: silica scaffolds in a powder state were deposited on a copper grid coated with a formvar/carbon membrane. Thermogravimetric analyses (TGA) were carried out under an oxygen flux (5 cm^3.min^{-1}) using a heating rate of 5°C.min^{-1}. The apparatus is a Stearam TAG-1750 thermo gravimetric analyser.

Heterogeneous catalysis. *Esterification batch process catalysis.* The reaction mixture contained 1-butanol (2 mmol), oleic acid (1 mmol), [C-*CRl*]@Glymo-Si(HIPE) (247 mg) and heptane (2 ml) in glass tubes. The reaction mixture was incubated at 37 °C during 24 hours. Production of ester was monitored using High Performance Liquid Chromatography (HPLC). *Transesterification batch process catalysis.* Transesterification reaction mixture contained trilinolein (100 μL of a 100 mg/mL previous solution in methyl tert-butyl ether), ethanol (25 mg), [C-*TLl*]@Glymo-Si(HIPE) (387 mg) and heptane (4 mL) in glass tubes. The reaction mixture was incubated at 37 °C during 24 hours. Conversion of trilinolein and production of ethyl linoleate were monitored using HPLC.

HPLC analytical system. The analytical system consisted of a 600 solvent delivery system, manuel injector (Waters, Milford, MA, USA). The compounds were separated on an Atlantis DC$_{18}$ (4.6mm×150mm, 5μm) column with an Atlantis DC$_{18}$ guard column (Waters). The column was operated at room temperature. Empower software (Waters) was used for data acquisition and processing. Standards were dissolved in tert-butyl methyl ether (MTBE). All solutions were filtered through a 0.45 μm membrane and degassed before use. The flow-rate was 1 mL/min and 20 μL of samples were injected. For the esterification catalyzed reactions a refractometer 410 (Waters, Milford, MA, USA) was used for detection. The mobile phase was acetonitrile (grade

HPLC), isocratic elution. For transesterification catalyzed reactions the detector in use was an ultraviolet diode array 996 (Waters, Milford, MA, USA). The maximum absorbance was at $\lambda =$ 204 nm. A gradient elution was used. Solvent A: acetonitrile, solvent B: tert butyl methyl ether. A: 4 min 100 (v/v) isocratic, A-B: 2 min 70/30 (v/v) gradient, A-B: 10 min 70/30 (v/v), A: 0.5 min 100 (v/v) gradient. Then, the column was equilibrated under above mentioned conditions for 10 min.

RESULTS AND DISCUSSION

Silica-based hybrid foams

In this study, enzymes have been confined within silica-based monolithic hybrid foams (Figure 1a) obtained through the use of a concentrated direct emulsion. Thus, the as-synthesized

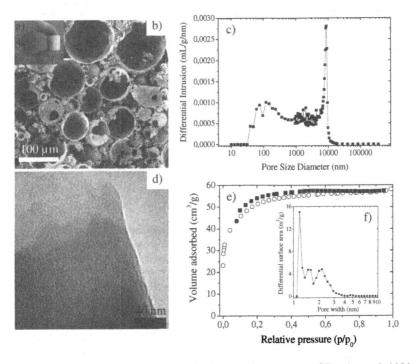

Figure 1. Lipase@Organo-Si(HIPE) characteristics. (a) Hybrid foam picture. (b) SEM micrograph. (c) Macroscopic pore size distributions obtained by mercury intrusion porosimetry. (d) TEM micrograph. (e) Nitrogen sorption curves (○ adsorption curve, ■ desorption curve). (f) Mesoscopic and microscopic pore sizes distribution obtained through the differential functional theory (DFT).

macroporous materials present a hollow spheres' aggregation structure (Figure 1b), due to the preferential mineralization at the oil/water interface. It's important to note the bimodal character

179

of the "window" sizes, centered at 100 nm and 10 μm (Figure 1c), checked by mercury intrusion porosimetry, reflecting the interconnected aspect of these open-cell monoliths. Consequently, these foams bear a high porosity, around 90 %. Moreover, the use of a lyotropic mesophase to stabilize the oil/water emulsion, induces the presence of vermicular-type mesostructure (Figure 1d). Consequently, the nitrogen sorption curves (Figure 1e) reveal a rapid rise in the low relative pressure, indicating microporosity, then, a weak hysteresis loop between the adsorption and desorption curves, associated with the presence of mesopores. This feature is confirmed by the pore sizes distribution, obtained through the density functional theory (DFT, Figure 1f), indicating mesopores centered at 2.2 nm. Nevertheless, due to the relatively small mesopores diameter, compared with the enzymes one (lipases are roughly spherical in shape, with a diameter around 4 nm), and the steric jamming of the post-grafted "Glymo" silane derivatives, it can be state that the enzymatic immobilizations occurred exclusively inside the macroporous network. The catalytic properties of this Lipase@Glymo-Si(HIPE) series have been checked, through esterification of fatty acids (using immobilized Candida Rugosa lipase; see Table I) and biodiesel production by transesterification of triacylglycerides (using immobilized Thermomyces Lanuginosus lipase; see Table I).

Table I. Hybrid foams Stoichiometries determined by both thermogravimetric and elemental analyses.

Lipase immobilized	Crude Candida Rugosa	Crude Thermomyces Lanuginosus
Stoichiometry	$[C-CRl]_{9.10-5}$ @$SiO_{1.91}(C_6O_2H_{11})_{0.09}$ • $0.09\ H_2O$	$[C-TLl]_{8.10-5}$ @$SiO_{1.95}(C_6O_2H_{11})_{0.05}$ • $0.06\ H_2O$
Enzymatic activity	$23.5\ \mu mol.min^{-1}.mg^{-1}$	$0.005\ \mu mol.min^{-1}.mg^{-1}$

Esterification batch process catalysis

The first reaction selected was the esterification of oleic acid with butanol to provide butyl oleate ester (Figure 2). This ester is an essential biodiesel additive acting as lubricant during winter use. As shown in Figure 2a, the immobilized catalyst was able to cycle for 19 runs reaching 100 % of conversion. The small decrease observed for the following runs can be attributed to the monolith partial collapse, leading to enzyme "leaching" into the reaction batch. Considering the fact that the immobilized biocatalyst has been kept at 4°C for two months, between cycles 10 and 11, catalytic performances are, in our knowledge, strongly enhanced, compared with immobilized enzyme-based esterification reported in literature [9]. Moreover, it's interesting to notice that the equivalent homogeneous catalysis of this esterification, using the same experimental set-up, reported an enzymatic activity 20 times lower than using the immobilized and confined lipase.

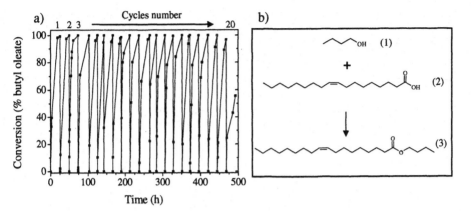

Figure 2. (a) Esterification reaction performances. (b) Reagents and product of the esterification: (1) butanol, (2) oleic acid, and (3) butyl oleate ester.

Transesterification batch process catalysis

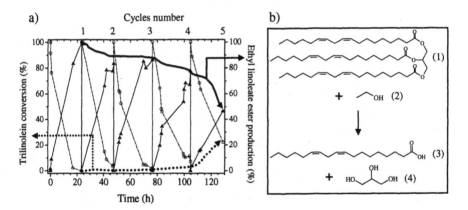

Figure 3. (a) Transesterification reaction performances. Each arrow refers to the y-coordinate indicated. (b) Reagents and products of the transesterification: (1) trilinolein, (2) ethanol, (3) ethyl linoleate ester, and (4) glycerol. *The intermediary products are not represented.*

Catalytic performances of these novel hybrid foams have also been checked through the transesterification reaction of trilinolein ester, which represents one of the olive oil major constituent, with ethanol (Figure 3). This reaction can be used to product biodiesel. As shown in Figure 3a, the first cycle allows reaching 100 % conversion within 24 hours at 37°C. Nevertheless, regarding the following runs, a constant decrease of the conversion can be

181

observed, probably due to the enzyme deactivation, owing to a very high ethanol concentration in the reaction media. In fact, in this study, a triglyceride/ethanol molar ratio of around 1/50 has been used, against 1/3 or 1/6 as a rule [10].

CONCLUSIONS

To conclude, the design of new enzyme-based monolithic catalysts and their catalytic performances over several reactions, have been successfully reached. In fact, this hybrid foams series combines both the advantages of macroporous networks, leading to low steric hindrance between biocatalysts and substrates, and covalent attachment, allowing good stability and, as a consequence, high cycling performances. Moreover, the inherent hydration of silica matrices (Table I) seems to be another advantage, while it is well known that water, in small amount, is an important feature toward optimizing transesterification biocatalytic performances [10]. The use of two different immobilized enzymes and reaction media (hydrolysis of triglyceride in water saturated non-aqueous medium has also been performed, but is not shown herein), shows an interesting versatility of the silica-based support. Last, but not least in view of industrial applications, the use of crude lipases represents a real economic asset.

Continuous flow heterogeneous biocatalysis, by using the same macrocellular supports, along with solvent minimization, are currently studied, and will be published in due course.

REFERENCES

1. A. M. Klibanov, Science 219, 722 (1983).
2. G. Zhou, Y. Chen, and S. Yan, Micro. Meso. Mater. 119, 223 (2009).
3. G. Mateo, G. Fernandez-Lorente, O. Abian, R. Fernandez-Lafuente, and J. M. Guisan, Biomacromolecules 1, 739 (2000).
4. M. T. Reetz, A. Zonta, and J. Simplekamp, Biotechnology and Bioengineering 49, 527 (1996).
5. R. Backov, Soft Matter 2, 452 (2006).
6. D. Barby, and Z. Haq, Eur. Patent 0060138 (1982).
7. N. Brun, B. Julian-Lopez, P. Hesemann, G. Laurent, H. Deleuze, C. Sanchez, M.-F. Achard, and R. Backov, Chem. Mater. 20, 7117 (2008).
8. N. Brun, A. Babeau-Garcia, C. Sanchez, and R. Backov, French Patent FR0954634 (2009).
9. R. Awang, M.R. Ghazuli, and M. Basri, Am. J. Biochem. Biotech. 3, 163 (2007).
10. N. Dizge, and B. Keskinler, Biomass and Bioenergy 32, 1274 (2008).

Mater. Res. Soc. Symp. Proc. Vol. 1272 © 2010 Materials Research Society   1272-PP09-14

Boundary Lubricant-Functionalized PVA Gels for Biotribological Applications

Michelle M. Blum[1] and Timothy C. Ovaert[1]
[1] University of Notre Dame, Notre Dame, IN 46556, USA

ABSTRACT

A novel material design was developed by functionalizing a biocompatible hydrogel material with organic boundary lubricants. Polyvinyl alcohol was functionalized with varying molar ratios of 0.2, 0.5, and 1.0 moles of lauroyl chloride. Tribological and mechanical characterization was performed by means of nanofriction testing and nanoindentation to determine the influence of the hydrocarbon chains on the friction coefficient and elastic modulus of the hydrogels. It was found that fusing of the lubricant to the polymer material has a positive effect on the surface friction properties, yet an unfavorable effect on stiffness properties of the gel due to the processing method.

INTRODUCTION

Hyaline cartilage is a material which exhibits ideal tribological properties by maintaining low friction which leads to high wear resistance between articulating joints. When damage to hyaline cartilage occurs, tissue regeneration is limited due to the avascular and aneural nature of cartilage. The resulting bone on bone contact causes pain and limited mobility. Current treatment options are limited to total or partial joint replacements, which have been successful, but are not ideal procedures due to long term failure of components and osteolysis. A vastly improved implant is desirable, utilizing materials that better mimic the structure and excellent tribological behavior of natural cartilage.

Hydrogels are a very promising option because they have similar properties to articular cartilage, including hydrated swelling, viscoelastic mechanical response and variable compressive stiffness. Therefore, they have the potential to tolerate multiple cycles of physiological loading. Considerable effort has gone into mechanical strengthening of hydrogels [1-3]. Yet, even with promising properties for strength mimicking, hydrogels exhibit poor friction properties. There have been a few experiments to investigate the improvement of hydrogel frictional and lubrication properties [3-5], but to achieve proper friction conditions, it is necessary to significantly improve the boundary lubrication capabilities of such bio-materials. A novel material design was developed by functionalizing a biocompatible hydrogel material with organic boundary lubricants.

EXPERIMENTAL DETAILS

2.1 Materials

Polyvinyl alcohol (PVA) with molecular weights (M_w) of 124 kDa to 186 kDa (Aldrich, 99 % hydrolyzed) was obtained from Sigma Aldrich. PVA was chosen for use in creation of a low friction hydrogel because it is biocompatible, permeable, easily prepared, and easily accessible. It also has a structure that can be easily modified to obtain desired properties. Lauroyl chloride (C12), a derivative of carboxylic acid, was obtained from Fisher Scientific. Carboxylic acids are derived from natural fats and oils, therefore they are environmentally friendly and biocompatible. Since the human body can metabolize fatty acids, when combined with a biomimetic hydrogel there will be little chance of adverse effects *in vivo*. Research has shown that natural lubricants which contain fatty acid concentrations produce excellent boundary lubrication properties [6].

2.2 PVA Functionalization

Functionalization occurs through a nucleophillic acyl substitution. In this reaction, the negatively charged chloride group which is bonded to the carbonyl group of the acid chloride increases the electronegative charge of the oxygen atom seen in Figure 1(a). The electron pair from the oxygen expels the chlorine ion, where the nucleophilic alcohol group from the PVA bonds easily to the carbonyl group carbon forming an ester bond, releasing a hydrochloric acid (HCl) molecule seen in Figure 1(b). The formation of (HCl) does not hinder the reaction because it was being produced in millimole concentrations. The final PVA with attached fatty acid is seen in Figure 1(c).

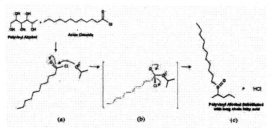

Figure 1: Diagram showing functionalization reaction.

2.3 Material Synthesis

The substituted PVA was prepared by dissolving approximately 30 millimoles of the polymer in 20 mL of the solvent N-methyl pyrrodine at 50°C [7]. The substituent, lauroyl chloride (C12), was then added into the solution in varying molar ratios of 0.2, 0.5, and 1.0 moles of substituent to moles of monomer. The complete solution was stirred for 24 hours at room temperature. Precipitation of the substituted polymer was obtained by supplementing the solution with DI water. The final solid was washed with DI water and dried under vacuum for 8 hours in order to ensure the excess water was removed.

2.4 Verification of Functionalization

In order to determine the composition of the reacted products a Fourier Transform Infrared (FT-IR) Spectrometer (Nicolet IR200, Thermo Electron Corporation, Waltham, MA) was used to collect atomic motion data about the compound. Samples were prepared for FT-IR analysis by combining the varying PVA/C12 precipitates with potassium bromide (KBr, FT-IR grade, Aldrich Chemical Company, Inc., Milwaukee, WI) in a 1:10 mass ratio and grinding the blend with a mortar and pestle. Pellets of the mixtures were formed in an evacuated die and then placed under a uniaxial pressure of 200 MPa with a hydraulic laboratory press [8]. Verification of functionalization was determined by observing the placement of the substituted hydrocarbons within the PVA spectra. Figure 2 shows the pure absorbance spectra of PVA with two distinct transmittance valleys; the alcohol groups visible as a wide blunt valley at wavelengths 3400-3650 cm^{-1} and alkyl bonds seen as two sharp valleys close together at 2850-2690 cm^{-1}, which was consistent with previous research [9]. Figure 2 also shows that after reaction, a clear ester valley was observed at 1735 cm^{-1} which verified a successful functionalization.

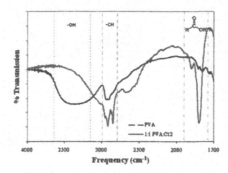

Figure 2: FT-IR results showing C12 chains bonded to PVA.

2.5 Fabrication of Hydrogels

Solutions of 30 wt% were prepared by dissolving either the pure PVA or the three varying molar concentrations of functionalized materials in a mixed solution with a ratio of 80:20 dimethyl sulfoxide (DMSO) and deionized (DI) water. The mixture was stirred for 60 minutes to allow for full dissolution of the polymer and homogenous mixing. The solutions were then heated at 90°C for 6 hours. The samples were subjected to five cycles of freezing and thawing during which the samples were frozen by placement in a commercial freezer regulated at -10°C for 20 hours. The samples were then removed from the freezer and allowed to thaw for 4 hours, and then returned to the freezer for another cycle. A fluid exchange procedure was performed in order to remove the DMSO solution from the gels. To fully hydrate the gels with DI water they were soaked in ethanol for 2 hours and then immersed in DI water for 72 hours, with the water changed every 24 hours. The hydrogel samples were produced as disks nominally 7 mm thick and 17 mm in diameter.

2.6 Tribological & Mechanical Characterization

In order to assess the influence of the hydrocarbon chains at the surface level, tribological and mechanical characterization was performed by means of nanofriction testing under boundary lubrication conditions and nanoindentation using a Triboindenter™ (Hysitron, Minneapolis, MN). For friction testing, a 100 µm radius diamond spherical tip was used as the sliding asperity. Constant normal loads of 300 and 500 µN were applied at a contact length of 10 microns producing a constant sliding speed of 333 nm/s. Four groups (Standard PVA, PVA-C12 5:1, 2:1, and 1:1) were tested. The average friction coefficient (µ) was calculated from ten randomized runs on each gel [10].

Nanoindentation tests were performed using a cylindrical stainless steel flat punch indenter with a diameter of 1000 µm and applied loads of 100 and 200 µN. Flat punch indentation has the advantage that the contact area between the indenter and the soft sample can be regarded as constant over the displacement range during testing. It is also beneficial during

185

testing since complete contact with the sample can be verified by a linear load vs. indentation depth curve. The elastic modulus (E) of five randomized indents on each gel was calculated using the Oliver-Pharr approximation which is applied to the unloading segment of a load-displacement indentation curve [11].

2.7 Equilibrium Water Content

Swelling properties of the gels were investigated by water content measurements. The mass of fully hydrated and dehydrated samples was measured. The water content (EWC) was calculated by the percent weight difference.

2.8 Contact Angle Measurements

Contact angle measurements were taken to quantify the hydrophobicity of the gels. The contact angle was measured by a goniometer (KRUSS G10, Germany). For this measurement a droplet of 18 megohm water was suspended from a micropipette on the surface of the gel. The angle was measured at 10 random locations on the sample.

RESULTS

Figure 3(a) shows the effect of C12 concentration on coefficient of friction. Comparison of frictional values between gel batches revealed that an increase in the concentration of lauroyl chloride led to a decrease in the coefficient of friction observed during transverse sliding. Standard PVA surfaces maintained reasonable values of $\mu = 0.16 \pm 0.09$ and $\mu = 0.15 \pm 0.06$ for loads of 300 and 500 μN respectively, while gels with ratio PVA-C12 of 1:1 saw a nominal decrease in μ of 67%, with values $\mu = 0.05 \pm 0.02$ and $\mu = 0.049 \pm 0.01$ for similar loading conditions. Figure 3(b) shows that elastic modulus values were on the order of kilopascals. An examination of the influence of load revealed that an increase in the molar concentration of lauroyl chloride led to a consistent decrease in compressive modulus. Standard PVA gels exhibited $E = 240.30 \pm 104$ kPa and $E = 342.47 \pm 170$ kPa for 100 and 200 μN indents, respectively, while the gels with the highest molar concentration (PVA-C12 1:1) had measured values of $E = 129.86 \pm 78$ kPa and $E = 124.18 \pm 71$ for similar loading conditions.

The contact pressures for some of the tests are summarized in Table I. At the interface of the contacting surfaces the pressure was calculated using simple Hertzian contact assumptions and material properties calculated from the indentation experiments. Loads 300 and 500 μN corresponded to pressures ranging from 93-172 kPa. Two-sample t-tests at the 95% confidence interval revealed that variation in load had no significant effect on friction coefficient or modulus. However, there was significant difference in the change in friction coefficient between standard PVA and gels with increasing C12 concentration above 0.5 molar ratios shown by Figure 3(a). A similar trend was observed for elastic modulus as seen in Figure 3(b).

The swelling ratio and contact angle of the gels were affected by the C12 concentration as well. Table I shows that as the concentration of C12 increased the water content increased and contact angle of the gel decreased. This occurred because the hydrocarbon chains hindered crystallization by increasing the distance between PVA chains, which decreased the effectiveness of van der Waals forces. Therefore, decreased bonding allowed for increased water absorption.

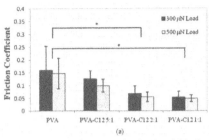
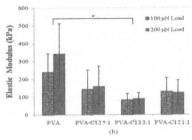

(a)   (b)

Figure 3: Results from mechanical and friction testing of hydrogels, (a) coefficient of friction and (b) elastic modulus.

DISCUSSION

An inverse correlation exists between coefficient of friction and C12 concentration ($r = -0.640$, $p < 0.002$), meaning that the fusing of the lubricant to the polymer material has a positive effect on the surface friction properties, yet increasing the lubricant concentration has an unfavorable effect on stiffness properties of the gel. Increasing the C12 content above 0.5 molar ratio decreased coefficient of friction significantly. The low friction produced could be attributed to formation of an interfacial water film due to the hydrophobicity of the hydrocarbon chains seen in Figure 4(a). Also, the free ends of the hydrocarbon chains could orient in a close packed formation due to cohesive forces, which provides a low shear strength film on the surface interface seen in Figure 4(b). Contact angle measurements were conducted to see if a close packed formation of hydrocarbon chains formed. If this occurred then the contact angle of the gel surface should have increased with increasing hydrocarbon concentration. Typical PVA contact angles range between 30 and 60 degrees based on M_w and PVA concentration within solution. Hydrophobicity has a direct relationship with the contact angle, namely a higher contact angle indicates higher hydrophobicity of the material [12]. The contact angle for the PVA hydrogels measured around 60 degrees. A slight increase in contact angle for the 2:1 batch was observed, but then a decrease in the 1:1 batch was found as shown in Table I. This inconsistent trend suggests that a hydrocarbon chain film did not form. It is thought that these angles were also affected by the water contained within the gel because there is an inverse relationship between contact angle and EWC. Increasing water content in the gel decreased the hydrophobicity of the total system. Performing water contact angle measurements of fully dehydrated gels will provide a better understanding of the surface properties exclusive to the matrix material.

Increases in lubricant concentration led to a decrease in the material modulus. This occurred because an increase in the concentration of functionalized bonds hindered the ability of the polymer to crosslink due to the opposite charge of the PVA and the hydrocarbon chains seen in Figure 4(c). This response is most likely a product of the processing method for the gels. A new processing method will be used and tested to see if a decrease in friction can occur while still maintaining a relatively stiff gel.

Functionalization of PVA was performed in an effort to improve the boundary lubrication properties of the material. Conjugation with simple carboxylic acid derivatives led to significant improvement of the tribological function of the biomaterial. This novel method provides an excellent platform for further investigation of synthetic biomaterials for tribological applications.

Table I: Equilibrium water content (EWC), contact angle and pressures for each gel batch.

Material	EWC (%)	θ (°)	Normal Load (μN)	Contact pressure (kPa)
PVA	59.0 ± 2.6	62.132 ± 6.2	300	145
			500	172
PVA-C12 5:1	55.5 ± 1.3	59.988 ± 2.7	300	125
			500	148
PVA-C12 2:1	77.9 ± 1.3	66.398 ± 4.2	300	78
			500	93
PVA-C12 1:1	72.3 ± 1.4	41.056 ± 1.9	300	93
			500	111

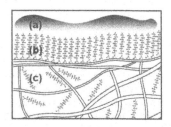

Figure 4: Possible lubrication mechanisms within the gel, including the (a) water layer, (b) hydrophobic C12 surface "film" and (c) entangled chains within the gel matrix.

REFERENCES

1. Thomas B.H., Fryman J.C., Liu K., Mason J.M., *Journal of the Mechanical Behavior of Biomedical Materials*, (2008), 2, p. 588-595.
2. Liu K., Ovaert .T.C., Mason J.J., *Journal of Material Science: Materials in Medicine*, (2008), 15(4): p. 1815-1821.
3. Gong J.P. and Osada Y., *Progress in Polymer Science*, 2002. 27, p. 3-38.
4. Nakashima K., Sawae Y., Murakami T., Tribology International, (2007). 40, p. 1423-1427.
5. Ma R., Xiong D., Miao F., Zhang J., Peng Y., *Materials Science and Engineering C*, (2009). p. 1979-1983.
6. Adhvaryu A., Biresaw G., Sharma B.K., Erhan S.Z., *Industrial &. Engineering Chemistry Research*, 2006, 45(10), p. 3735-3740.
7. Orienti, I, Zuccari G., Luppi B., Zecchi V. *Journal of Microencapsulation*, 2001, 18(1), p. 77-87.
8. Roeder R.K., Converse G., Leng H., Yue W., *Journal of the American Ceramic Society*, 89(7), 2096-2104 (2006).
9. Miranda,T.M.R., Gonc A.R., and Amorim M.T.P.. *Polymer International*, 2001, 50, p. 1068-1072.
10. Johnson K. L., *Contact Mechanics* (Cambridge University Press, Cambridge, 1985).
11. Oliver, W.C. and G.M. Pharr, *Journal of Materials Research*, 1992. 7(6), p. 1564-1583.
12. Thongphud A., Paosawatyanyong B., Visal-athaphand P. and Supaphol P. *Advanced Materials Research* Vols. 55-57 (2008), p. 625-628.

Mater. Res. Soc. Symp. Proc. Vol. 1272 © 2010 Materials Research Society   1272-PP04-03

A Critical Assessment of RNA-Mediated Materials Synthesis

Stefan Franzen[1] and Donovan Leonard[2,3]
[1] Department of Chemistry, North Carolina State University, Raleigh, NC, 27695, U.S.A.
[2] Mat. Sci. & Tech. Division, Oak Ridge Natl. Laboratory, Oak Ridge, TN, 37831-6376, U.S.A.
[3] Dept. of Materials Sci. & Eng., University of Tennessee, Knoxville, TN, 37996-2200, U.S.A.

ABSTRACT

RNA- and DNA-mediation or templating of materials has been used to synthesize nanometer scale wires, and CdS nanoparticles. However, RNA and DNA have the potential to act as catalysts, which could be valuable tools in the search for new routes to materials synthesis. RNA has the ability to catalyze splicing and cutting of other RNA molecules. Catalytic activity has been extended to more general classes of reactions for both RNA and DNA using *in vitro* selection methods. However, catalytic activity in materials synthesis is a more recent idea that has not yet found great application. The first example of RNA-mediated evolutionary materials synthesis is discussed with specific data examples that show incompatibility of reagents in the solvent system utilized. The hydrophobic reagent $Pd_2(DBA)_3$, used as a metal precursor, was observed to spontaneously form nanostructures composed of $Pd_2(DBA)_3$ or $Pd(DBA)_3$ rather than palladium nanoparticles, as originally reported [1]. A case study of this materials synthesis example is described including the complimentary use of multi-length scale techniques including transmission electron microscopy (TEM), selected area electron diffraction (SAED), scanning TEM (STEM), electron energy loss spectroscopy (EELS), scanning electron microscopy (SEM), energy dispersive X-ray spectroscopy (EDS) and optical microscopy (OM). This example raises important questions regarding the extent to which non-aqueous solvents should be used in nucleic acid-mediated processes, the nature of selections in enzyme and materials development, and the requirement for chemical compatibility of the precursor molecules. The importance of good characterization tools at every stage of an *in vitro* selection is illustrated with concrete examples given. In order to look at the way forward for nucleic acid-mediated materials synthesis, an examination of the chemical interaction of nucleic acids with various precursors is considered. Application of density functional theory calculations provides one means to predict reactivity and compatibility. The repertoire of chemical interactions in the nucleic acids is considered vis-à-vis common metals and metal chalcogenides. The case is made for the need for water-soluble syntheses and well-controlled kinetics in order to achieve the control that is theoretically possible using nucleic-acids as a synthetic tool.

INTRODUCTION

The concept of RNA catalysis has revolutionized modern biology. RNA is not only a messenger that carries a transcript of the information contained in DNA, but it is also an enzyme capable of processing other RNAs in a number of ways. In the spliceosome, small nuclear RNAs (snRNAs) combine to splice out non-coding sequences known as introns. RNase P is a ribonucleoprotein capable of splicing the ends of transfer RNA (tRNA). There are two important features of RNA regulation and catalysis; nucleobase stacking/pairing/triplex interactions and the nucleophilicity of the 2'-OH. In fact, the 2'-OH group of the ribose sugar is the only difference between RNA and DNA. The reactivity of the remaining parts of RNA and DNA is exceptionally low. This is likely by design, since high reactivity would lead to side reactions that could corrupt the genetic code. In the nucleobases the "reactive" groups are the exocyclic amino and the imino group. Attempts to make artificial enzymes of RNA have met with some success, particularly where the hydrophobic effect is important for catalysis. These considerations lead to the conclusion that RNA catalysis in nature plays a significant role in RNA regulation, but not a general role in catalytic processes of life. Indeed, more recent considerations suggest that the "RNA world" hypothesis should be thought of as a "RNA-protein world" hypothesis, and the key point may be that RNA preceded DNA [2].

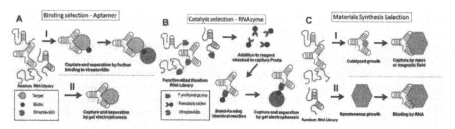

Figure 1. Three types of *in vitro* selection for random RNA libraries.

RNA selections for binding interactions shown in Figure 1A have been known since 1990 [3,4]. For a RNA binding sequence or aptamer, the selection is based on capture by the target followed by separation from non-binding sequences. In designing artificial selections for non-natural enzymatic catalysis or materials synthesis using RNA, shown in Figures 1B and 1C, respectively, the limited repertoire of functional groups has led to the concept of non-native functional groups attached to particular nucleobases. The most commonly modified RNA base is uridine, which has a natural modification at the 5-position that inspired synthetic modification at that position. Modified uridine triphosphates in the 5-position act as substrates for RNA polymerases, such as T7 RNA polymerase so that they may be incorporated into a sequence provided by a DNA template and can be used to make RNA libraries of $\sim 10^{14}$ sequences for the selection [5].

The concept of using modified RNA to synthesize new materials was first put into practice in 2004 using a pyridyl modification (pyRNA) and a precursor that contains Pd, tris(dibenzylideneacetone) dipalladium(0) or $Pd_2(dba)_3$, with the goal of mediation of metal-metal bond formation and synthesis of controlled morphology Pd nanoparticles [1]. Hexagonal particles were observed in studies of the precipitate from the solutions containing 1 μM pyRNA and 100 μM $Pd_2(dba)_3$ [1]. Subsequently, cubic particles were reported that were synthesized by a distinct sequence from the hexagonal particles [6]. However, as shown in Figure 1C, appropriate care must be taken to ensure that the selection involves synthesis by the RNA (mechanism I) and not spontaneous particle formation followed by binding (mechanism II). Herein, we examine the composition of the particles, their properties and the conditions used in the selection in order to better understand these results.

EXPERIMENT

We have examined the solvent conditions required for the selection by testing the solubility of $Pd_2(dba)_3$ in mixtures of tetrahydrofuran (THF) and water used in the selection [7]. It is known that $Pd_2(dba)_3$ is completely insoluble in water. Yet, Ref. 1 states that the selection was done in aqueous solution without qualification. That value was later revised to 5% THF in water [6]. However, an Erratum gives the value as 50% THF [8], whereas a more recent Correction states that a range from 1-10% THF was used (Figure 2A)[9]. Since there are conflicting values, we conducted a quantitative study of the solubility of $Pd_2(dba)_3$ in THF/water mixtures provided the following fit to the Flory-Huggins model for solubility in mixed solvents shown in Figure 2C. Starting with the standard free energies of formation of solutions of $Pd_2(dba)_3$ in pure H_2O and THF, which are $\Delta G_{S,w}^o$ and $\Delta G_{S,THF}^o$, respectively, the free energy for solution in the mixture, ΔG_S, can be expressed using Flory-Huggins theory:

$$\Delta G_S^o/RT = \phi_w(\Delta G_{S,w}^o/RT) + \phi_{THF}(\Delta G_{S,THF}^o/RT) + \phi_w \ln \phi_w + \phi_{THF} \ln \phi_{THF} + \chi_{WTHF} \phi_w \phi_{THF} \quad (1)$$

where $\phi_w = V_w/V$ and $\phi_{THF} = V_{THF}/V$ are the volume fractions of water and THF, respectively and χ_{WTHF} is the Flory-Huggins interaction parameter for THF and H_2O. By knowing the solubility in THF we can calculate $\ln a_1 = -\Delta G_S^o/RT$ in the mixed solvent system. Given estimates of $\Delta G_{S,THF}^o = 12.2$ kJ/mol and $\Delta G_{S,w}^o = 42.5$ kJ/mol, based on our experimental data, combined with an estimate of 1.5 for the Flory-Huggins parameter of a THF/H_2O mixture, we can calculate the solubility plot. One important point that derives from this analysis is that the solubility changes in an exponential, rather than a linear, fashion as a function of the volume fractions.

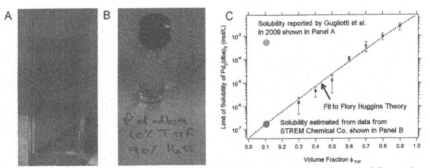

Figure 2. Solubility studies of $Pd_2(dba)_3$. A.) Picture from E-letter to Science and Correction stating that cuvette contains 400 µM $Pd_2(dba)_3$ in 10% THF/90% H_2O [1,9] B.) Experimental result by STREM Chemical Co. of 400 µM $Pd_2(dba)_3$ in 10% THF/90% H_2O after filtration C.) Quantitative solubility measurement, with dots to show agreement of 2B, but not 2A.

We have examined the synthesis and chemical composition of hexagons reported to be produced by RNA mediation. The hexagons showed the following properties:

1. They do not require pyRNA for formation, but form spontaneously in THF/H$_2$O mixtures [7-10,11].

2. Hexagons, formed on a Au slide, dissolve in pure THF, as observed by OM (Figure 3)[7].

A   B

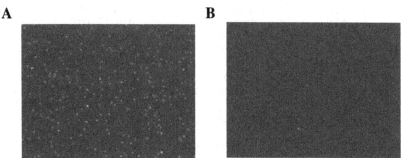

Figure 3. Dissolution of hexagonal particles in pure THF. A.) Hexagonal particles formed on a Au surface. B.) The same surface after 30 second exposure to the pure THF. Appropriate care was taken to avoid mechanical disturbance of the particles. Data were obtained using a Zeiss LSM-5 Pascal Microsope.

3. Based on both SEM/EDS and STEM/EELS analysis the hexagons were found to be composed of 95% carbon [7-10,11].

4. Electron diffraction showed stages of beam damage (Figure 4)[7]. These are successively
(stage 1) a spot pattern consistent with the unit cell parameter of Pd$_2$(dba)$_3$ [t < 20 seconds]
(stage 2) subsequent degradation of the pattern to an amorphous pattern [t < 3 minutes]
(stage 3) faint rings appear, which can be indexed to Pd metal [t > 5 minutes]

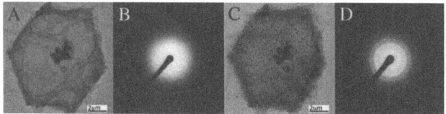

Figure 4. A.) TEM micrograph and B.) diffuse SAED indicative of a hexagon that has lost crystallinity (stage 2) C.) TEM and D.) SAED with an increase in intensity of a ring pattern due to 2-3 nm Pd nanoparticles formed by 5 minute exposure to the 200 kV electron beam (stage 3)[7].

5. The hexagons can grow to large size and the X-ray crystal structure determined to show that they consist of Pd(dba)$_3$, i.e. a Pd atom lost from the original Pd$_2$(dba)$_3$.

We examined the range of particle morphologies found in evaporated droplets of Pd$_2$(dba)$_3$/THF solution by OM, SEM and TEM/STEM. The structures formed on a substrate after evaporation of a 100% THF solvent droplet included stars, rods, hexagons and cubes.

192

Regular hexagons are formed at > 70% THF solution, irregular hexagons in the range from 30-60% THF and mostly spherical Pd nanoaprticles (200 nm in diameter) were observed at 10% THF. Crystalline hexagons were observed with both red and yellowish-brown color. The former are most likely $Pd_2(dba)_3$ and the latter were determined to be $Pd(dba)_3$ by X-ray crystallography.

DISCUSSION

The concept of using modified RNA for materials synthesis is potentially useful if the properties of RNA are properly considered. Since RNA is soluble in water, the selections should be done in aqueous solvent. Even small amounts of non-aqueous solvent can modify the structure of the RNA. It is evident from Ref. 8 and from the solubility studies in Figure 2 that >50% THF was used in the selection and syntheses. Such high concentrations of THF will likely denature the RNA. Nonetheless, the THF/H_2O mixed solvent used by the authors of Ref. 1 is consistent with the observation that the hexagonal particles form in the absence of RNA. The fact that the RNA is likely denatured under the conditions actually used in Ref. 1 is of no consequence since RNA is not required for the formation of the hexagons.

The ability of metal ions to cleave RNA must be carefully considered in the selection. The mechanism of RNA cleavage by the 2'-OH is catalyzed by divalent ions. While Mg^{2+} is the native ion most frequently involved, other metal ions such as Pb^{2+}, Co^{2+}, etc. can significantly accelerate RNA cleavage. This limits the concentration of these ions. Using precursors that are water-soluble and appropriate ionic conditions, we examine the conditions needed to select for products of RNA-mediated evolutionary materials synthesis.

Selection criteria for materials synthesis

Artificial selection is challenging. Natural selection is based only on survival of an organism so that it can reproduce. Artificial or *in vitro* selection must merely separate the active sequences of RNA from inactive ones. The selection of aptamers, which bind to target proteins or other analytes, is relatively straightforward (Figure 1A)[3,4]. In that case the sequences that bind are immobilized. If the target is tagged with biotin, it can be easily separated by binding to avidin on a surface, a magnetic bead, or for separation by gel electrophoresis. A selection for enzymatic activity is more challenging. In that case, that RNA must catalyze a reaction that leads to bond formation in such a way that the product can bind to avidin, i.e. it contains a biotin. This requires that there be two reactants, one attached (tethered) to the RNA and a second reactant that is attached to biotin (Figure 1B). Materials synthesis requires the RNA to assemble components (precursors) into the new material and then remain bound to the product to permit selection and separation based on mass difference (Figure 1C). Since the selection itself involves one or at most a few RNAs of similar sequence that have the capability to synthesize the material, the process must be catalytic in order to proceed. For example, a hexagon that is 500 nm across and 50 nm thick would consist of ~9,000 atoms of Pd. Since Pd-Pd bonds would be formed as the dba ligand is stripped off, the RNA would have to carry out a repetitive process. This is unlike templating that allows a material to form in a particular shape or with a particular crystal habit based on the presentation of a macromolecular surface. The word mediation is not clear since it has been used either to mean both catalysis and templating in reference to the same process [1,6]. Finally, one must consider the sensitivity of RNA to degradation by impurities or by metal ions used in the selection.

Materials characterization

The example of *in vitro* selection for synthesis of Pd relied mainly on high spatial resolution electron microscopy as a means to validate the selection. TEM/STEM and SEM are useful for multi-length scale determination of particle size and shape, but do not provide chemical composition or atomic ordering (e.g. crystalline or amorphous) unless combined with the methods of EELS, EDS and/or SAED. EELS data have been reported to achieve atomic column sensitivity and typically provides sub-nanometer resolution for examining single nanoparticles [12]. SAED can sample areas on the order of ~1 μm and provides information on whether the material in the selected area is a single crystal (spot pattern), polycrystalline (ring pattern) or amorphous (diffuse rings). If a suitable standard is used to calibrate the TEM camera length, sub-Ångström level lattice parameters values are obtainable with SAED. The initial validation using only TEM[1,6] led to the hypothesis that the hexagons were composed of Pd metal. Subsequent analysis using EDS, EELS, SAED, OM[7-10,11] and X-ray crystallography proved that the particles are not Pd, but are composed of molecular crystals of $Pd_2(dba)_3$ and the degradation product $Pd(dba)_3$ that we have observed by X-ray crystallography.

CONCLUSIONS

There is no evidence that crystalline Pd metal particles were formed by RNA mediation in any of the hexagons studied. Analysis shows that the hexagons formed in THF/H_2O solutions are composed of the metal precursor, $Pd_2(dba)_3$ and $Pd(dba)_3$, a decomposition product. The precipitation observed in control experiments, when THF solution of $Pd_2(dba)_3$ is mixed with water, shows that particle formation is rapid compared to any RNA-mediated process. These studies do not rule out a role for RNA in materials synthesis in general. Rather these results suggest conditions (aqueous, low divalent ion, and water soluble-precursors) and combined methods of analysis (TEM, EELS, EDS, and SAED) that may be used in future attempts to harness the power of *in vitro* selection for discovery of novel nanomaterials.

REFERENCES

(1) L. A. Gugliotti, D. L. Feldheim, and B. E. Eaton, *Science* **304**, 850 (2004).
(2) T. R. Cech, *Cell* **136**, 599 (2009).
(3) C. Tuerk, and L. Gold, *Science* **249**, 505 (1990).
(4) A. D. Ellington, and J. W. Szostak, *Nature* **346**, 818 (1990).
(5) T. M. Tarasow, S. L. Tarasow,and B. E. Eaton, *Nature* **389**, 54 (1997).
(6) L. A. Gugliotti, D. L. Feldheim, and B. E. Eaton, *J. Am. Chem. Soc.* **127**, 17814 (2005).
(7) S. Franzen, M. Cerruti, D.N. Leonard, and G. Duscher, *J. Am. Chem. Soc.* **129**, 15340 (2007).
(8) S. W. Chung, A. D. Presely, S. Elhadj, S. Hok, S. S. Hah, A. A. Chernov, M. B. Francis, B. E. Eaton, D. L. Feldheim, and J. J. De Yoreo, *Scanning* **30**, 474 (2008).
(9) L. A. Gugliotti, D. L. Feldheim, and B. E. Eaton, *J. Am. Chem. Soc.* **131**, 11634 (2009).
(10) D. N. Leonard, M. Cerruti, G. Duscher, and S. Franzen, *Langmuir* **24**, 7803 (2008).
(11) D. N. Leonard, and S. Franzen, *J. Phys. Chem. C* **113**, 12706 (2009).
(12) S. J. Pennycook, M. Varela, A. R. Lupini, M. P. Oxley, and M. F. Chisholm, *J. Elecron. Micro.* **58**, 87 (2009).

Mater. Res. Soc. Symp. Proc. Vol. 1272 © 2010 Materials Research Society   1272-PP04-13

Direct atomistic simulation of brittle-to-ductile transition in silicon single crystals

Dipanjan Sen[1,2], Alan Cohen[2,3], Aidan P. Thompson[4], Adri C.T. van Duin[5], William A. Goddard III[6], Markus J. Buehler[2,*]

[1] Department of Materials Science and Engineering, Massachusetts Institute of Technology, 77 Mass. Ave., Cambridge, MA 02139, USA
[2] Laboratory for Atomistic and Molecular Mechanics, Department of Civil and Environmental Engineering, Massachusetts Institute of Technology, 77 Mass. Ave. Room 1-235A&B, Cambridge, MA,02139, USA
[3] Department of Mechanical Engineering, Massachusetts Institute of Technology, 77 Mass. Ave., Cambridge, MA 02139, USA
[4] Multiscale Dynamic Materials Modeling Department, Sandia National Laboratories, PO Box 5800, MS 1322, Albuquerque, NM, 87185, USA
[5] Department of Mechanical and Nuclear Engineering, Pennsylvania State University, University Park, PA 16802, USA
[6] Division of Chemistry and Chemical Engineering, California Institute of Technology, 1201 E. California Blvd., Pasadena, CA 91125, CA, USA

* Corresponding author, Electronic address: E-mail: mbuehler@MIT.EDU

ABSTRACT

Silicon is an important material not only for semiconductor applications, but also for the development of novel bioinspired and biomimicking materials and structures or drug delivery systems in the context of nanomedicine. For these applications, a thorough understanding of the fracture behavior of the material is critical. In this paper we address this issue by investigating a fundamental issue of the mechanical properties of silicon, its behavior under extreme mechanical loading. Earlier experimental work has shown that at low temperatures, silicon is a brittle material that fractures catastrophically like glass once the applied load exceeds a threshold value. At elevated temperatures, however, the behavior of silicon is ductile. This brittle-to-ductile transition (BDT) has been observed in many experimental studies of single crystals of silicon. However, the mechanisms that lead to this change in behavior remain questionable, and the atomic-scale phenomena are unknown. Here we report for the first time the direct atomistic simulation of the nucleation of dislocations from a crack tip in silicon only due to an increase of the temperature, using large-scale atomistic simulation with the first principles based ReaxFF force field. By raising the temperature in a computational experiment with otherwise identical boundary conditions, we show that the material response changes from brittle cracking to emission of a dislocation at the crack tip, representing evidence for a potential mechanisms of dislocation mediated ductility in silicon.

INTRODUCTION

The mechanical response of solids subject to extreme applied stress is controlled by atomistic mechanisms in the vicinity of stress concentrations such as crack tips (Figure 1(a)) [1-3]. Crack tips represent mathematical singularities for the stress distribution, providing local large interatomic forces that form the seeds for macroscopic failure [1, 3-6]. In brittle materials, the material responds by further extension and resulting growth of cracks, leading to a catastrophic failure through fragmentation. In ductile materials, the repeated shear of lattice planes through dislocations leads to macroscopic permanent change of the shape of the material without catastrophic failure [1, 3, 4, 7]. These two extreme cases of brittle and ductile material response are summarized schematically in Figure 1(b). Whether a material is ductile or brittle depends on the competition of intrinsic material parameters (such as the energy required to create new surfaces, versus the energy required to initiate shearing of a lattice to form a dislocation). Alternatively, it has been shown that the type of failure response can be controlled by the temperature. Low temperatures tend to lead to more brittle, and higher temperature to a more ductile material response [8, 9]. In silicon, experimental studies of single crystals with a crack have shown that the material response can be either brittle or ductile, depending on the temperature [10]. At temperatures below ≈800 K silicon tends to be extremely brittle, while it exhibits ductile behaviour above ≈800 K [10]. This surprising phenomenon has been studied experimentally and theoretically for more than 20 years. Several explanations have been proposed. Rice's criteria for dislocation nucleation at crack tips [4] describe BDT as a dislocation nucleation-controlled event [11]. Others describe the BDT in terms of dislocation mobility-controlled events [12, 13], along with changes in the density of dislocations emitted [14].

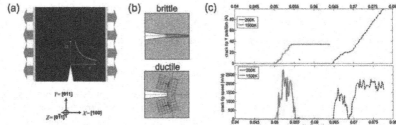

Figure 1: Subplot (a): Silicon single crystal under mode I loading with edge crack of {100} < 110 > character. Subplot (b): Schematic illustration of brittle (crack extension) versus ductile material (dislocation nucleation) behaviour. Subplot (c): Crack tip position and velocity for low (200 K) and high (1500 K) temperatures, as a function of engineering strain $\varepsilon = \Delta L_X / L_{0,X}$. Crack blunting and stopping accompanied by dislocation emission is observed at higher temperature, whereas at 200 K brittle cleavage fracture is observed. An equilibrium crack speed of ≈2,000 m/s is achieved at 200 K. The 200 K simulation results show a temporary slowdown of the crack due to a sessile 5-7 defect in the crack pathway [15] (at approximately 6.9% strain or 34.5 ps). The formation of the 5-7 defect lowers the stress intensity at the crack tip by bond-rotations, and slows the crack to an arrest, before the increasing loading allows the crack to proceed again. Several more such arrest phenomena occurrences may be expected in longer simulations over larger crack paths due to the thermally-activated nature of these defects.

The development of atomistic level understanding of BDT has been hindered partly due to the lack of atomistic models that enable the simulation of sufficiently large systems to accurately describe the fracture processes associated with fracture at a range of temperatures. For silicon, these processes involve several tens of thousands of atoms surrounding a crack tip for time periods of fractions of nanoseconds. Describing bond breaking processes in silicon has

196

required quantum mechanical (QM) methods to properly describe the complex electronic rearrangements that determine the barriers and hence the rates, where large changes in bond angles and coordination can affect the interatomic forces [15-18]. However, accurate QM calculations for large system sizes that would be required to describe the complex details of bond rearrangements under large stresses are currently impractical. An alternative approach has been to apply relatively simple empirical relationships between bond stretch and force [6, 19, 20], but earlier results suggested that fracture in silicon cannot be modelled with such force fields [17].

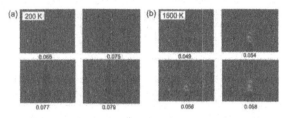

Figure 2: Subplot (a): Crack motion at 200 K. The numbers indicate engineering strain values at the particular snapshots. Brittle cleavage fracture with smooth surfaces is observed. Subplot (b): Crack motion at 1500 K. Slight crack opening followed by sudden crack blunting with emission of a dislocation is observed. (Note; in subplot (b) we use a different visualization scheme than in subplot (a). To better visualize the shear of atomic planes, we keep the initial bond connectivity, which leads to the plotting of overstretched bonds that easily indicate the slip plane. In subplot (a) we use dynamic bond redrawing in VMD.)

METHODS

We apply the first principles based ReaxFF reactive force field, which retains nearly the accuracy of QM, even for bond breaking events. The ReaxFF parameters are determined solely by fitting to QM-data on silicon and silicon oxide chemistry [21]. The ReaxFF computational costs are only one magnitude higher than ordinary force fields, enabling parallelized applications to systems larger than 1,000,000 atoms at nanosecond time-scales [22]. ReaxFF therefore allows us to directly simulate the BDT without any fitting against experimental results or fracture properties. Earlier studies with the ReaxFF force field have shown that it reproduces key experimental observations of purely brittle fracture of silicon [15, 18]. Figure 1(a) shows the simulation geometry, a single crystal of silicon with a surface crack under mode I tensile loading. We carry out two simulations, one at 200 K and one at 1500 K, with exactly the same initial and boundary conditions (other than the temperature) and the same constant loading rate throughout the simulation (increase of loading is never stopped). We use the GRASP simulation code (General Reactive Atomistic Simulation Program), a new code developed to carry out the large-scale molecular dynamics simulations of crack growth using reactive force fields. GRASP was developed at Sandia National Laboratories, Albuquerque, New Mexico. GRASP uses a spatial decomposition algorithm [23] in order to distribute the calculation of atom positions, velocities and forces over a large number of processors. The ReaxFF charge equilibration equations are solved every time step using a sparse parallel conjugate gradient algorithm [24]. This approach provides excellent parallel scaling on general-purpose computing clusters. Typical MD simulations using the ReaxFF force field using GRASP can be run efficiently on up to $N/200$ processors, where N is the number of atoms. Beyond this point, the parallel efficiency drops below 50%. One MD time step with the ReaxFF force field requires about 10^{-3} processor-seconds per atom, making it considerably more expensive than conventional force fields such as

Tersoff and Stillinger-Weber. We consider a perfect silicon crystal oriented so that the x-y-z directions are $(1,0,0) \times (0,1,1) \times (0,\overline{1},1)$, with an initial crack on a (100) fracture plane with initial [011] fracture direction. The slab geometry used is approximately 20 nm by 18 nm, with an initial crack length of 4 nm. We use periodic boundary conditions in the out-of-plane direction, imposing purely in-plane strain in a plane-strain setup. The thickness of the systems is 15 Å, and the system consists of ~27,500 atoms. To apply load, we continuously strain the system according to mode I by displacing the boundaries [6, 19] at a strain rate of 2×10^9 s^{-1} (see Figure 1(a)). The system evolution is under a canonical ensemble, with temperature being controlled by a Berendsen thermostat. Loading is initiated after a system equilibration of 50,000 MD steps at the target temperature. Runs at each temperature are repeated two times.

RESULTS AND DISCUSSION

Figure 1(c) shows the results of the crack tip history and the crack tip velocity history, respectively, for both temperatures. A significant difference in the behaviour can be observed. In the 200 K case, fracture initiates at approximately 6.5% applied strain (or 32.5 ps) and the crack speed quickly approaches 2 km/sec, in agreement with earlier simulations [15, 17] and experimental results [25]. For the 1500 K case, a small regime of crack extension initiates at 5% applied strain (or 25 ps), but the crack arrests quickly and does not propagate any further. The initial load of 5% is slightly smaller than the load for the low temperature, which is attributed to thermal activation effects. The high-temperature discontinuation of crack propagation, despite the increase of applied loading, suggests that an alternative mechanism is activated that is responsible for dissipating the energy supplied through continued elastic loading at the boundary of the slab.

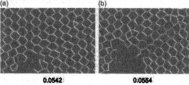

Figure 3: Subplot (a-b): Details of atomic structure close to crack tip around the time of dislocation emission in the 1500 K simulation. (a), viewing into the <110> direction shows the crack just before partial emission, with distorted hexagonal silicon cages (numbers indicate engineering strain values). (b) shows the characteristic atomic structure of a stacking fault on the glide plane (outlined in red) showing that a partial dislocation has passed.

The analysis of the atomistic structure during the deformation process explains this observation. Figure 2(a) displays snapshots of crack dynamics at 200 K, showing clean brittle fracture through generation of almost perfectly flat atomic surfaces. The analysis at 1500 K shown in Figure 2(b) reveals a drastically different behaviour, showing the nucleation of a partial dislocation from the crack tip. This observation demonstrates clearly the change of the response of silicon from brittle to ductile, solely due to a rise in the temperature, and confirms that the BDT is associated with a distinct change in the atomistic deformation mechanism. Immediately preceding either crack extension or dislocation nucleation, both systems behave virtually identically, showing no signature of either response. Figure 3 provides further detailed views into the region at the crack tip for the case of 1500 K. Figure 3(b) displays the crack tip configuration with a nucleated dislocation, showing the existence of a small stacking fault region and the dislocation core. The resistance to crack extension is determined by the fracture surface energy, describing the amount of energy needed to create new material surfaces during crack growth [4, 5, 26]. The resistance to nucleation of dislocations is described by a similar parameter, the unstable stacking fault energy (USF), describing the energy barrier associated with creating a dislocation in a crystal. There are two possible slip planes, the "shuffle plane" and the "glide plane" [27, 28]. The USF value for the shuffle plane is lower than that of a glide-plane partial. However, our simulation results show that a partial dislocation is emitted despite the higher energy barrier. This observation agrees with results from experimental studies of this phenomenon [29]. Figure 4(a) shows another snapshot of the region close to the crack tip after the dislocation has been nucleated at 1500 K. We find that a 90° partial dislocation is emitted on a ($\bar{1}$11) glide plane, with the slip direction [211]. Figure 4(b) shows the structure of the incipient dislocation, where the incipient dislocation is highlighted by a red ellipse. The incipient dislocation appears extended and its size extends over 4 hexagonal silicon rings, viewed along the <110> direction. The formation of this wide incipient dislocation, which evolves into a glissile partial, requires a large activation energy, and it could be that it is only favoured at high temperatures. Notably, at all temperature, sessile 5-7 crystal defect formation is observed at the crack tip [15]. However, whereas at low temperature this defect remains sessile (and can lead to crack tip deflection and instability; see [15]), at higher temperature this defect provides a seed for formation of the incipient stage of the glissile partial dislocation.

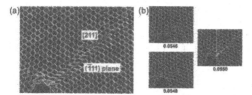

Figure 4: Subplot (a): Dislocation emission at crack tip at 1500 K. A 90° partial dislocation is emitted on a ($\bar{1}$11) glide plane and the crack stops growing (slip direction [211]). Subplot (b): Structure of the incipient dislocation at 1500 K (position outlined in a red ellipse). The numbers below each snapshot indicate engineering strain. Dislocation emission occurs at approximately 5.46% strain, or 27.3 ps.

The results reported here provide the first atomistic-scale view of the BDT in silicon. The use of a full chemistry full physics, first principles based atomistic model in the simulations is essential to directly simulate BDT. In earlier simulation models we used a hybrid formulation [15, 18] that involved only up to 3,000 reactive ReaxFF atoms in the crack tip vicinity. We observe that whereas this smaller reactive region is sufficient to describe brittle fracture of silicon, it is too small to capture the formation of dislocations from the crack tip at elevated temperatures. This is explained by the observations shown in Figures 3 and 4 that reveal extended disordered regions. A reactive description of these domains, which requires more than 20,000 atoms, is crucial in order to reflect the transition to ductile behaviour. After the success of phenomenological dislocation models in describing BDT [4, 14, 26, 30], progress has been somewhat limited. This may have been partly due to the absence of an atomistic simulation model that was capable of capturing the chemistry of silicon bond breaking and thus the atomistic mechanisms of BDT. Our results may enable the development of first principles based dislocation mechanics models. Since ReaxFF is capable of describing a diverse range of materials, the approach used here for silicon as a brittle "model material", provides a practical means to studying the coupling of chemical reactions to mechanical properties in other brittle materials. The results reported here leave many questions open, including a more detailed analysis of the mechanisms that result in the transition from brittle cracking to dislocation emission, the study of a more gradual increase of the temperature (and the investigation of the transition occurs suddenly as observed in experiment, with an identification of the actual transition temperature), and a careful analysis of the rate dependence of the transition temperature. An enhanced understanding of the mechanical properties of silicon could enable us to utilize silicon in the development of novel bioinspired functional nanomaterials, as the large-scale manufacture of silicon nanostructures is possible with current technology derived from semiconductor manufacturing. These and other issues could be addressed in future work.

ACKNOWLEDGMENTS: The authors acknowledge support from ARO (W911NF-06-1-0291), DARPA and NSF-ITR. AC acknowledges support from MIT's UROP program.

REFERENCES

[1]   L. B. Freund, *Dynamic Fracture Mechanics* (Cambridge Univ. Press, 1990).
[2]   K. B. Broberg, *Cracks and Fracture* (Academic Press, 1990).
[3]   J. P. Hirth, and J. Lothe, *Theory of Dislocations* (Wiley-Interscience, 1982).
[4]   J. R. Rice, and R. M. Thomson, Phil. Mag. **29**, 73 (1974).
[5]   J. R. Rice, J. Mech. Phys. Solids **40**, 239 (1992).
[6]   M. J. Buehler, and H. Gao, Nature **439**, 307 (2006).
[7]   M. J. Buehler, *Atomistic modeling of materials failure* (Springer (New York), 2008).
[8]   P. Gumbsch et al., Science **282**, 1293 (1998).
[9]   A. Strachan, T. Cagin, and W. A. Goddard, J. Comp.-Aided Mat. Des. **8**, 151 (2002).
[10]   C. S. John, Philosophical Magazine **32**, 1193 (1975).
[11]   M. Khantha, and V. Vitek, Acta Materialia **45**, 4675 (1997).
[12]   P. B. Hirsch, S. G. Roberts, and J. Samuels, Proceedings of the Royal Society of London. Series A, Mathematical and Physical Sciences (1934-1990) **421**, 25 (1989).
[13]   A. Hartmaier, and P. Gumbsch, Physica status solidi. B. Basic research **202**, 1 (1997).
[14]   Y. B. Xin, and K. J. Hsia, Acta Materialia **45**, 1747 (1997).
[15]   M. J. Buehler et al., Phys. Rev. Lett. **99**, 165502 (2007).
[16]   R. D. Deegan et al., Phys. Rev. E **67**, 066209 (2003).
[17]   N. Bernstein, and D. W. Hess, Physical Review Letters **91**, 025501 (2003).
[18]   M. J. Buehler, A. C. T. van Duin, and W. A. Goddard Iii, Physical review letters **96**, 95505 (2006).
[19]   M. J. Buehler, F. F. Abraham, and H. Gao, Nature **426**, 141 (2003).
[20]   D. Holland, and M. Marder, Phys. Rev. Lett. **80**, 746 (1998).
[21]   A. C. T. v. Duin et al., J. Phys. Chem. A **107**, 3803 (2003).
[22]   K. I. Nomura et al., Physical Review Letters **99** (2007).
[23]   S. Plimpton, Journal of Computational Physics **117**, 1 (1995).
[24]   M. B. R. Barrett, et al. *Templates for the Solution of Linear Systems: Building Blocks for Iterative Methods* (1994).
[25]   J. A. Hauch et al., Phys. Rev. Lett. **82**, 3823 (1999).
[26]   J. R. Rice, and G. B. Beltz, J. Mech. Phys. Solids **42**, 333 (1994).
[27]   M. S. Duesbery, and B. Joos, Philosophical Magazine Letters **74**, 253 (1996).
[28]   Y. M. Juan, and E. Kaxiras, Philosophical Magazine A **74**, 1367 (1996).
[29]   S. W. Chiang, C. B. Carter, and D. L. Kohlstedt, Phil. Magazine A **42**, 103 (1980).
[30]   P. B. Hirsch, and S. G. Roberts, Philosophical Magazine A **64**, 55 (1991).

Mater. Res. Soc. Symp. Proc. Vol. 1272 © 2010 Materials Research Society   1272-PP06-03

Biopreparation of Highly Dispersed Pd Nanoparticles on Bacterial Cell and Their Catalytic Activity for Polymer Electrolyte Fuel Cell

Takashi Ogi, Ryuichi Honda, Koshiro Tamaoki, Norizo Saito, Yasuhiro Konishi
Chemical engineering, Osaka Prefecture University, Sakai, 599-8531, Japan.

ABSTRACT

Rapid development in the area of low-temperature fuel cells has led to increased attention on catalyst synthesis with cost effective and environmentally-benign technology (green chemistry). In this study, a highly dispersed palladium nanoparticle catalyst was successfully prepared on a bacterial cell support by a single-step, room-temperature microbial method without dispersing agents. The metal ion reducing bacterium *Shewanella oneidensis* were able to reduce palladium ions into insoluble palladium at room temperature when formate was provided as the electron donor. The prepared biomass-supported palladium nanoparticles were characterized for their catalytic activity as anodes in polymer electric membrane fuel cell for power production. The maximum power generation of the biomass-supported palladium catalyst was up to 90% of that of a commercial palladium catalyst.

INTRODUCTION

Low-temperature fuel cells generate power by direct electrochemical conversion of fuel (hydrogen/methanol) with oxidant (oxygen/air) to produce water/CO_2. In a fuel cell, the anode (fuel electrode) and cathode (oxidant electrode) are installed on either side of a polymer electrolyte membrane. Each electrode is coated on one side with a thin catalyst layer. Hydrogen/methanol is fed into the anode side of the fuel cell and oxygen/air enters through the cathode side. In the case of hydrogen, it is dissociated by the catalyst on the anode and transformed into a proton and an electron. The electrons and protons flow through an external circuit and the polymer electrolyte membrane, respectively. They then react with oxygen to form water at the cathode.

The catalyst used in a fuel cell is composed of noble metal nanoparticles and support materials. Supported metal nanoparticle catalysts have been prepared by various methods including microwave irradiation [1-3], chemical vapor deposition [4], impregnation and reduction of metal precursors in a microporous support [5,6], the colloidal method [7,8], and the microemulsion method [9,10]. These methods generally require an elevated temperature to complete the reduction of soluble noble metals and expensive protecting agents such as surfactants to inhibit agglomeration of the nanoparticles and the support material. Any organic stabilizer must then be removed from the catalyst often by heating at a high temperature. Therefore, it is essential to develop a novel nanoparticle catalyst preparation method that uses less toxic precursors (e.g. water as the solvent), fewer reagents and synthetic steps, and a reaction temperature close to room temperature.

Microbial methods are attractive for the direct preparation of noble metal nanoparticle catalysts on bacterial cells as they are economical, safe, and environmentally friendly [11-18]. Windt et al. successfully prepared Pd nanoparticle using the metal ion-reducing bacterium

Shewanella oneidensis with H$_2$ or formate as electron donor [13]. Furthermore, they applied its *S.oneidensis* supported Pd nanoparticle to catalysts in a dehalogenation of polychlorinated biphenyls [13], lindane [16], and trichloroethylene [18]. However, the application of the *S.oneidensis* supported Pd to fuel cell electrode catalysis has not been reported.

Here, we report the biosynthesis of a high performance biomass-supported palladium catalyst for fuel cell application. We successfully deposited highly dispersed palladium(0) nanoparticles on bacterial cells by a single-step, room-temperature microbial reduction using the metal ion-reducing bacterium *Shewanella oneidensis*. In the evaluation of catalytic activity, the biomass-supported palladium was directly used for testing in a fuel cell without heating.

EXPERIMENT
Bacterial strain and growth conditions

S. oneidensis ATCC 700550 was obtained from the American Type Culture Collection (ATCC). The bacterial strain was grown aerobically in trypticase soy broth medium at 30 °C and pH 7.2. After 10–12 h of batch inoculation the *S. oneidensis* cells were harvested by centrifugation, re-suspended in KH$_2$PO$_4$/KOH buffer (pH 7.0), and pelleted again by centrifugation. This procedure was repeated twice, and the washed cells were subsequently re-suspended in KH$_2$PO$_4$/KOH buffer (pH 7.0). The cell suspension was bubbled with N$_2$ for 30 min and immediately used for the microbial reduction of palladium(II).

Preparation of palladium nanoparticles on the bacterial cells

An anaerobic glovebox was used to carry out microbial reduction experiments. In a typical reduction experiment at 25 °C, 5 mL of *S. oneidensis* cell suspension was added to 10 mL of aqueous PdCl$_2$ solution under anaerobic conditions with a gas phase of N$_2$. The mixed solution was buffered at pH 7.0 with KH$_2$PO$_4$/KOH (pH 7). The cell concentration in the mixed solution was held at (6.3±0.4)×10^{15} cells/m^3. The initial concentration of PdCl$_2$ ranged from 1 to 20 mol/m^3. Sodium formate was provided as the electron donor for microbial reduction. The number of *S. oneidensis* cells in the solution was counted in a Petroff-Hausser counting Chamber with a microscope. The *S. oneidensis* cells and biomass-supported particles were observed by transmission electron microscopy (TEM) using a JEOL model JEM-2100F equipped with an EDX attachment (JEOL model JED-2300T). X-ray Absorption Near Edge Structure (XANES) was used to determine the oxidation states of palladium in the *S. oneidensis* cells after exposure to aqueous PdCl$_2$ solution at pH 7.0 for 120 min in the presence of formate. Reference samples were palladium foil (Pd(0)) and 10 mol/m^3 aqueous PdCl$_2$ solution. Measurements were conducted at the BL14B2 line in the Japan Synchrotron Radiation Research Institute (JASRI) in Hyogo, Japan.

Evaluation of catalytic activity in a fuel cell

The biomass-supported palladium nanoparticles were dried and tested for their catalytic activity as anodes in a PEMFC for power production. After microbial reduction tests, the palladium nanoparticles on the biomass were harvested by centrifugation, washed with deionized water, and dried at 50 °C for 6 h. Before performing a fuel cell test, the palladium content of the dried biomass was accurately determined by completely dissolving the sample in aqua regia for quantitative analysis using inductively-coupled plasma (ICP) spectroscopy. In a typical fuel cell test using a PEMFC Kit (Techno Xpress Inc.), the biomass-supported palladium catalyst for the anode inks was prepared by mixing the dried biomass-supported palladium nanoparticles with carbon particles, 10 wt% nafion solution (1.0 mL), and ultra pure water (0.4 mL). Commercial platinum catalyst (Chemix. Co. Ltd.) was used as the cathode ink for all the tests. In the preparation of the membrane electrode assembly (MEA), anode and cathode inks were uniformly printed onto a nafion membrane to give a palladium loading of 1.28 mg/cm^2 on the anode and 0.16 mg/cm^2 on the cathode for the H$_2$/O$_2$ PEMFC test. After drying the anode and cathode inks, each catalyst layer was attached to a piece of Teflon-treated carbon paper (Toyo Corporation). The active area of a single cell was 6.25 cm^2. For the PEMFC test, hydrogen gas was supplied to the anode at 300 mL/min. Current (I) and voltage (V) were measured and recorded against resistance (R) from 0.1–200 Ω. A commercial palladium catalyst (Pd/C) was obtained from Wako Pure Chemical Industries Ltd. for comparison of electrical properties.

DISCUSSION

When the formate was provided as the electron donor, the PdCl$_2$ solution including the *S. oneidensis* cells changed color from pale yellow to dark brown. The appearance of dark brown provided a convenient visible signature for the microbial formation of palladium nanoparticles. However, in a sterile control medium, the PdCl$_2$ solution without formate or *S. oneidensis* cells didn't change the color. Thus, the observed change in the PdCl$_2$ solution color was presumably caused by the microbial reduction of soluble Pd(II) into insoluble palladium with the electron donor.

Figure 1(a) shows a typical TEM image of palladium nanoparticles prepared by *S. oneidensis* cells with 40 mol/m^3 formate. *S. oneidensis* is a rod-shaped bacterium approximately 0.5 μm wide and 2 μm long. When the *S. oneidensis* cells were exposed to PdCl$_2$ solution, a number of highly dispersed nanoparticles were observed on the bacterial cells. These results suggest the biomass-supported palladium nanoparticles will likely be suitable for application as palladium catalysts. From the high magnification TEM image (Fig. 1 (b)), the nanoparticles were determined to be 5–10 nm in primary particle size.

XANES was used to determine the oxidation state of the biomass-supported palladium nanoparticles. Figs. 1 (c) shows the XANES spectrum of palladium in the *S. oneidensis* cells after exposure to 1 mol/m^3 aqueous PdCl$_2$, together with XANES spectra for reference compounds of aqueous PdCl$_2$ (Pd(II)) and metallic palladium (Pd(0)). The bacterial samples and palladium foil had identical XANES spectra, which demonstrates that *S. oneidensis* cells could reduce palladium(II) ions into elemental palladium.

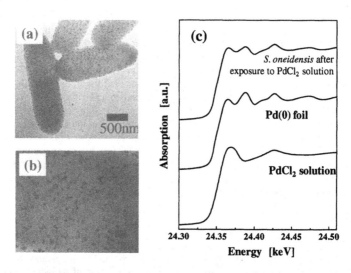

Figure 1 Low (a) and high (b) magnification TEM images of palladium nanoparticles prepared by *S. oneidensis* cells at 25 °C. c) XANES spectra of palladium in the *S. oneidensis* cells, in metallic Pd foil, and in a 10 mol/m^3 aqueous PdCl$_2$ solution.

Effect of initial palladium(II) concentration

The microbial reduction was carried out at various initial palladium(II) concentrations (5–20 mol/m^3) to investigate the effects on particle size and dispersity. In these experiments, the initial formate concentration was held at 50 mol/m^3. Figure 2 shows the TEM images of biomass-supported palladium nanoparticles prepared using various initial palladium(II) concentrations with a reaction time of 120 min. The TEM images show that agglomerated particles were observed at the lower initial palladium(II) concentration. An increase in the initial palladium(II) concentration to 10 and 20 mol/m^3 resulted in the formation of fine, homogeneous, and separated nanoparticles on most bacterial cells (Figs. 2(b) and (c)). The geometric mean diameter d_p and geometric standard deviation σ_g were evaluated by direct counting of approximately 1000 nanoparticles per sample from TEM photographs using image analysis software (Ruler). For nanoparticles prepared at initial palladium(II) concentrations of 10 (Fig. 2(d)) and 20 mol/m^3 (Fig. 2(e)), these were d_p=5.26 nm, σ_g=1.35 and d_p=7.31 nm, σ_g=1.38, respectively. The highest surface area of 75.7 m^2/g was attained when the initial palladium(II) concentration was 10 mol/m^3. These results illustrate that the initial palladium(II) concentration in the starting solution is important during preparation of highly dispersed palladium nanoparticles on bacterial cells.

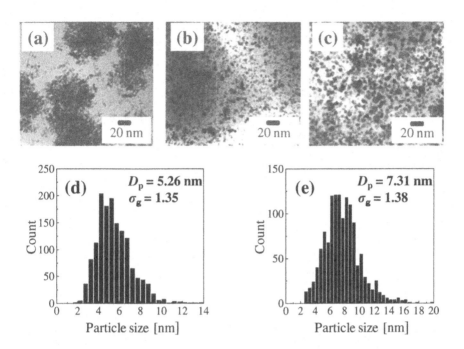

Figure 2 TEM images and particle size distribution of palladium nanoparticles prepared at different initial palladium(II) concentrations using *S. oneidensis* cells: a) 5 mol/m^3, b) 10 mol/m^3, c) 20 mol/m^3, d) 10 mol/m^3, and e) 20 mol/m^3.

Catalytic activity of biomass-supported palladium

Four biomass-supported palladium catalysts were prepared using different initial palladium(II) concentrations in the starting solution. The catalytic activities of these were examined on application as anode electrode catalysts in PEMFCs. All tests were repeated three times using separate batches of the biomass-supported palladium catalysts. Figure 3(a) shows the effect of biomass-supported palladium specific surface area on maximum power density of the PEMFC. All values are the average of the three separate experiments. The dried biomass-supported palladium prepared at 10 mol/m^3 generated the maximum power density of 4.78 mW/cm^2. The dried biocatalyst prepared at 5 mol/m^3 exhibited a peak power density of 0.63 mW/cm^2, which is about 7.5 times smaller than that of the dried biocatalyst prepared at 10 mol/m^3. Furthermore, the power density was correlated with the specific surface area of biomass-supported palladium. Increases in the power density appear to occur as a result of increases in the catalytic surface area of the well-dispersed biomass-supported palladium nanoparticles.

For comparison, the electrical properties of chemically synthesized commercial carbon-supported palladium nanoparticles (Pd/C) were measured under the same conditions as the biomass-supported palladium catalyst (Fig. 3(b)). Despite the low-temperature synthesis, the maximum power density of the dried biocatalyst prepared at 10 mol/m^3 was up to 90 % that of the commercial Pd/C catalyst. These results demonstrate that the prepared biomass-supported palladium catalyst exhibits good electrical properties within a fuel cell without heating.

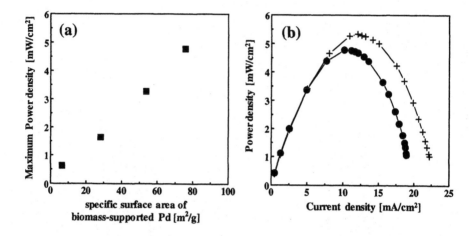

Figure 3 a) the effect of specific surface area of biomass-supported palladium on maximum power density of PEMFC, b) I-P curves of PEMFC using different types of palladium anode catalyst: (●) the dried biomass-supported palladium, and (+) commercial palladium catalyst.

CONCLUSIONS

Highly dispersed palladium nanoparticles were successfully produced on the cell surface of *S. oneidensis* at room temperature when formate was provided as the electron donor. The initial palladium(II) concentration is an important factor in the microbial method for controlling the particle size and dispersity on the bacterial surface. XANES analysis revealed that resting cells of *S. oneidensis* are able to reduce palladium(II) ions into elemental palladium(0). The maximum power density of the highly dispersed palladium nanoparticles was up to 90 % of that of a commercial palladium nanoparticle catalyst. Our methodology for producing highly dispersed palladium nanoparticles on a bacterial cell support is an attractive green process that is economical, environmentally benign, and can produce high efficiency catalysts.

ACKNOWLEDGMENTS

This work was supported by a Grant-in-Aid for Young Scientists (Start-up) (21810025) from the Ministry of Education, Science, Sports and Culture, Japan. Parts of this research were performed with the approval of the Japan Synchrotron Radiation Research Institute (JASRI) in Hyogo, Japan (No. 2008A1789). We thank Dr. H. Ofuchi (JASRI, Hyogo, Japan) for assistance with the XANES study.

REFERENCES

1. G. Glaspell, L. Fuoco, and M.S. El-Shall, *J. Phys. Chem. B* **109**, 17350 (2005).
2. G. Glaspell, H.M.A. Hassan, A. Elzatahry, V. Abdalsayed, and M.S. El-Shall, *Top. Catal.* **47** 22, (2008).
3. J.M. Campelo, T.D. Conesa, M.J. Gracia, M.J. Jurado, R. Luque, J.M. Marinas, and A.A. Romero, *Green Chem.* **10**, 853 (2008).
4. M. Okumura, S. Nakamura, S. Tsubota, T. Nakamura, M. Azuma, and M. Haruta, *Catal. Lett.* **51**, 53 (1998).
5. M. Watanabe, M. Uchida, and S. Motoo, *J. Electroanal. Chem.* **229**, 395 (1987).
6. A.Barau, V.Budarin, A. Caragheorgheopol, R.L uque, D. J. Macquarrie, A. Prelle, V. S. Teodorescu, and M. Zaharescu, *Catal. Lett.* **124**, 204 (2008).
7. C.H. Liang, W. Xia, H. Soltani-Ahmadi, O. Schluter, R.A. Fischer, and M. Muhler, *Chem. Commun.* **2**, 282 (2005).
8. Z.H. Zhou, S.L. Wang, W.J. Zhou, G.X. Wang, L.H. Jiang, W.Z. Li, S.Q. Song, J.G. Liu, G.Q. Sun, and Q. Xin, *Chem. Commun.* **3**, 394 (2003).
9. Z.L. Liu, J.Y. Lee, M. Han, W.X. Chen, and L.M. Gan, *J. Mater. Chem.* **12**, 2453 (2002).
10. X. Zhang and K.Y. Chan, *Chem. Mat.* **15**, 451 (2003).
11. V. Baxter-Plant, I. P. Mikheenko, L. E. Macaskie, Biodegradation **14**, 83 (2003).
12. V. S. Baxter-Plant, I. P. Mikheenko, M. Robson, S. J. Harrad, L. E. Macaskie, Biotechnology Letters **26** 1885 (2004).
13. W. De Windt, P. Aelterman, W. Verstraete, Environmental Microbiology **7** 314 (2005).
14. W. De Windt, N. Boon, J. Van den Bulcke, L. Rubberecht, F. Prata, J. Mast, T. Hennebel, W. Verstraete, Antonie Van Leeuwenhoek International Journal of General and Molecular Microbiology **90** 377 (2006).
15. S. Harrad, M. Robson, S. Hazrati, V. S. Baxter-Plant, K. Deplanche, M. D. Redwood, L. E. Macaskie, Journal of Environmental Monitoring **9** 314 (2007).
16. B. Mertens, C. Blothe, K. Windey, W. De Windt, W. Verstraete, Chemosphere **66** 99 (2007).
17. M. D. Redwood, K. Deplanche, V. S. Baxter-Plant, L. E. Macaskie, Biotechnology and Bioengineering **99** 1045 (2008).
18. T. Hennebel, H. Simoen, W. De Windt, M. Verloo, N. Boon, W. Verstraete, Biotechnology and Bioengineering **102** 995 (2009).

Mater. Res. Soc. Symp. Proc. Vol. 1272 © 2010 Materials Research Society   1272-PP06-12

Assembly of Ag@Au Nanoparticles Using Complementary Stranded DNA Molecules and Their Detection Using UV-Vis and Raman Spectroscopic Techniques.

Derrick Mott, Nguyen T. B. Thuy, Yoshiya Aoki, and Shinya Maenosono.
School of Materials Science, Japan Advanced Institute of Science and Technology, 1-1 Asahidai, Nomi, Ishikawa 923-1292, Japan

ABSTRACT

Silver nanoparticles coated by a layer of gold (Ag@Au) have received much attention because of their potential application as ultra sensitive probes for the detection of biologically important molecules such as DNA, proteins, amino acids and many others. However, the ability to control the size, shape, and monodispersity of the Ag@Au structure has met with limited success. In our own research we have addressed this challenge by creating an aqueous wet chemical synthesis technique towards size and shape controllable Ag@Au nanoparticles. These materials are highly interesting because of the tunable silver core size, and the tunable gold shell thickness, opening many avenues to the modification of the particle properties in terms of bio-molecular sensing. The resulting nanoparticle probes were functionalized with two complementary stranded DNA oligonucleotides. When combined, the complementary strands hybridized, causing the Ag@Au nanoparticles to assemble into large nano-structures. The presence of the oligonucleotide was confirmed through a series of techniques including UV-Vis and Raman spectroscopy, as well as TEM, XPS, DLS, and many others. The results reflect the role that the nanoparticle physical properties play in the detection of the bio-molecules, as well as elucidate the characteristics of the bio-molecule/nanoparticle interaction.

INTRODUCTION

To date there have been several attempts to detect DNA using nanoparticles as sensitive probes [1]. Such detection techniques often rely on assembly of complementary stranded NPs (Ag, Au or Ag@Au NPs) and their subsequent hybridization which is detected by colorimetric techniques or by labeling of the nanoparticles with a reporter label which can then be detected using Raman Spectroscopy [1]. These techniques open the doors to the ultra-sensitive detection of DNA using nanoparticle probes, but there is still much work to be done in understanding the nanoparticle probe properties themselves. Both silver and gold nanoparticles (NPs) have received wide attention for their enhanced properties in a multitude of potential applications such as sensing, microelectronics, and catalysis, [2,3] because of the many desirable chemical and physical properties of the materials. In terms of bio-diagnostics and sensing, it is the optical properties that make Ag NPs exceptional, while for Au NPs it is the resistance to oxidation and enhanced thiol chemistry that is attractive [1,4]. The current trend in this area of research is the coupling of these two materials as a core@shell structure that takes advantage of the optical properties of silver and the stability/thiol chemistry of gold. These Ag@Au nanoparticles are expected to have use as bio-probes with unprecedented sensitivity and selectivity. Despite the excitement surrounding these materials though, there are still many challenges to address, including the ability to synthesize Ag NPs in aqueous phase with a desired size, shape, or monodispersity [5], and the ability to coat the silver particle with a uniform and controllable layer of gold. Finally, the detection of DNA using these nanoscale probes presents many unique challenges of its own including attachment of DNA to the nanoparticle probe surface, assembly of the NPs using a target strand of DNA, hybridization of the matched DNA strands, and finally detection of the DNA [6]. Our approach to these unique challenges is a straightforward assembly of the Ag@Au nanoparticle probes using specific sequenced DNA strands through a photoligation reaction, and finally detection of the interaction using a dual buffer/reporter molecule with Raman spectroscopy.

EXPERIMENT

Chemicals. Silver nitrate, sodium acrylate, gold tetrachloroaurate trihydrate and common solvents were obtained from Aldrich. Water was purified with a Millipore Direct-Q system (18.2 MΩ). Dialysis membranes with molecular weight pore size of 10,000 daltons were obtained from Spectra/Por and were rinsed in pure water before use. Calcium chloride and sodium chloride were purchased from Wako Chemicals. Cacodylate buffer solution (0.2M) was purchased from Nacalai Tesque Inc., Japan. Nuclease P1 was purchased from Yamasa Corp., Japan. ODN sequences were synthesized using an Applied Biosystems 3400 DNA synthesizer. The reagents for the synthesis of DNA components was purchased from Glen Research, USA.

Synthesis of Ag NP Cores: Ag NPs were synthesized by first mixing 50ml of water with 1.25×10^{-5} moles of silver nitrate, and then adding 6.75×10^{-6} moles of sodium hydroxide, which results in a dilute yellow colored solution of silver hydroxide. This solution is purged with argon and is then brought to reflux. At reflux, 2.55×10^{-4} moles of sodium acrylate are added causing the solution to turn completely clear. The solution is refluxed for 1 hour, over this time the solution color changes from clear to green-yellow to yellow-orange.

Purification of as-synthesized Ag NPs: Prior to deposition of Au on the Ag NP seeds, the as-synthesized particles are purified to remove excess acrylate, silver, sodium, and other ions from the solution. Purification is performed by enveloping the particle solution inside of a cellulose dialysis membrane with pore size of 10,000 daltons and soaking in a distilled water bath. The water was changed every 12 hours for 48 hours.

Deposition of Au on the Ag Cores to form Ag@Au NPs: 50ml of the dialysized Ag particles are brought to reflux and 10ml of a gold tetrachloroaurate trihydrate solution (ranging from 6.25×10^{-7} to 3.13×10^{-6} moles according to the thickness of the Au shell desired) and 10ml of a sodium acrylate solution (from 5.10×10^{-5} to 2.55×10^{-4} moles) are added dropwise simultaneously. The solution color changes depending on the amount of Au added. In general, as Au and sodium acrylate is added to the Ag particles, the color changes from yellow-amber to dark amber to grey to grey-purple and finally to purple.

Instrumentation and Measurements: An array of techniques including Transmission Electron Microscopy (TEM), High Resolution TEM and Energy Dispersive Spectroscopy (HR-TEM, EDS), X-Ray Photoelectron Spectroscopy (XPS) and Ultra-Violet Visible Spectroscopy (UV-Vis) were used to characterize the size, shape, composition and other properties of the materials. TEM analysis was performed on an Hitachi H-7100 transmission electron microscope operated at 100kV. HR-TEM and EDS analysis was performed on an Hitachi H-9000NAR transmission electron microscope operated at 300kV. TEM samples were prepared by dropping the suspended particles onto a carbon coated copper grid and drying in air overnight.

DISCUSSION

In general, our synthetic rout towards Ag@Au NPs consists of 2 main steps. First we synthesize the Ag cores in aqueous phase with an acrylate capping agent. In the first part of the results and discussion section we show our synthetic approach to monodispersed Ag particles. The second step of this research is the coating of Au on the Ag particle surface. In the second part of the results and discussion section we illustrate the coating of the Ag particles by Au and discuss the ability to control the thickness of the Au layer and the resulting morphology of the nanostructures. The overall technique of our synthetic approach is shown in Scheme 1. In this approach, first we used silver nitrate and sodium hydroxide to create a dilute solution of silver hydroxide. This solution was brought to reflux, whereupon sodium acrylate was added, initiating the formation of Ag NPs. The Ag particles formed over the course of one hour as evidenced by the appearance of a yellow-amber solution. Figure 1A shows the TEM image of the as-synthesized Ag NPs capped by the acrylate molecule. The particle size distribution is 20.5 ± 3.3 nm (16% deviation). Given the size of the particles and their total mass, we calculated the Ag NP concentration to be 7.28×10^{-11} M and their extinction coefficient to be 2.11×10^{10} $M^{-1}cm^{-1}$.

After synthesis, the particles were purified by using dialysis to remove excess acrylate, silver ions, and other species. The coating of the silver NPs by gold is achieved in what is essentially a seeded growth reaction. Briefly, first the Ag NPs are brought to reflux, then dilute aqueous gold and sodium acrylate solutions are added simultaneously, dropwise, causing the Au to reduce at the Ag NP surface, causing a coating of Au to be formed. The primary challenge in this reaction is the propensity of Ag to be oxidized as Au is reduced during the coating. Such an occurrence can lead to either hollow Au particles or alloyed Au and Ag NPs. In our own reaction, the alloying or etching of the Ag NPs can be minimized through control of the concentration of gold being added, the rate that it is added, and by addition of the sodium acrylate capping and reducing agent. Here, the role of sodium acrylate is to cause the Au to become reduced before the Ag can be oxidized.

Scheme 1: Reaction Rout for the Synthesis of Ag NPs and their Coating with Au.

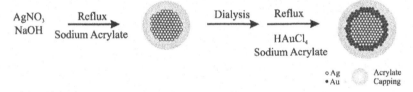

Figure 1 shows a series of TEM images of Ag@Au NPs synthesized using the method described above. For each sample, the amount of gold and sodium acrylate added was incrementally increased. Figure 1B shows a sample of Ag@Au NPs synthesized by adding 6.25×10^{-7} moles of Au and 1.38×10^{-4} moles of sodium acrylate (corresponding to 5% Au in terms of atomic composition). Inspection of the TEM image reveals particles with a uniform spherical morphology. The direct observation of a deposition of a layer of gold on the surface of the silver nanoparticles is difficult to ascertain in this image. The size distribution of these particles is 17.5 ± 3.7nm. Figure 1C shows a sample of Ag@Au NPs synthesized by adding 1.88×10^{-6} moles of Au and 1.38×10^{-4} moles of sodium acrylate (corresponding to 15% Au in terms of atomic composition). The TEM image shows several particles with a spherical morphology, but now several particles are observed that have a lighter spherical center and darker outside ring. The observation of this dark outside and light center is inconsistent among different particles in the sample, some particles display no darker ring and light center at all, or some particles display simply a light spot near the periphery of the NP. We attribute this observation to the formation of an incomplete Au shell on the Ag particle surface. In effect, the round holes that are observed on the particles in the TEM image are a gap or hole in the Au shell, allowing us to see the Ag core inside the particle. The size distribution of these NPs is 16.3 ± 2.7 nm. In addition, a few NPs are observed in the TEM image with much smaller size than the parent Ag NPs (~9nm), which could be attributed to the non-specific formation of Au NPs without coating on the Ag surface. The presence of these small particles was not observed in techniques such as UV-Vis, probably because of their low concentration as compared to the coated particles. Finally, Figure 1D shows a sample of Ag@Au NPs synthesized by adding 3.13×10^{-6} moles of Au and 1.38×10^{-4} moles of sodium acrylate (corresponding to 25% Au in terms of atomic composition). The TEM image reveals many particles with a light center and thick dark outside. Now the particles seem to have adopted roughly hexagonal or pentagonal shapes, likely reflecting the tendency of Ag nanocrystals to be oriented in the twinned structure, templating the growth of Au at their surface. The size distribution of these particles is 17.5 ± 5.1nm. Among these three samples the particle sizes are generally smaller than the precursor Ag NP seeds (size of 20.5 ± 3.3 nm). We attribute this size decrease to a small degree of etching of the silver surface at the initial reaction stage as the gold layer is deposited. As the reaction progresses, the acrylate reducing agent plays a more significant role in reducing the gold as it is deposited on the particle surface, thereby preventing the entire silver core from being etched away.

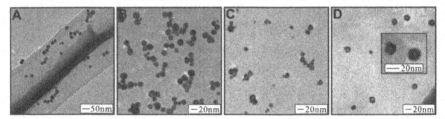

Figure 1. TEM images for as-synthesized Ag NPs (A), and Ag@Au NPs with atomic feeding ratio of: 5% Au (B), 15% Au (C), and 25% Au (D).

The nanoparticles synthesized above were used as probes for the detection of DNA (Ag@Au NPs synthesized with 5% Au in terms of atomic feeding ratio). Scheme 2 shows our approach to the detection of DNA. Scheme 2A shows the components of the detection. The DNA sensing technique exploits the photoligation reaction between oligodeoxynucleotides (ODNs) attached on the surfaces of NPs in the presence of target DNA (T-DNA). In this approach, two ODNs, a common probe (P1) and a specific discriminating probe (P2) are separately conjugated on the surface of the NPS (Scheme 2B). The conjugated NPs hybridize and form aggregates via photoligation between the P1 and P2 strands (Scheme 2B and C). Once aggregation has taken place the interparticle spacing between the NPs acts as a "hot spot", causing a SERS signal from the Raman active buffer (cacodylic acid) used in the assembly.

Scheme 2: Assembly Scheme for Ag@Au NPs using DNA in a Photoligation Reaction.

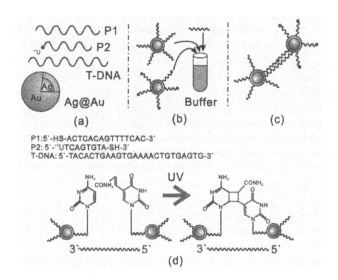

P1:5'-HS-ACTCACAGTTTTCAC-3'
P2: 5'-"UTCAGTGTA-SH-3'
T-DNA: 5'-TACACTGAAGTGAAAACTGTGAGTG-3'

Figure 2 shows Raman spectra for the functionalization of the Ag@Au NPs with the P1, P2 and target DNA strands. Figure 2a shows the Raman spectra of the sample before incubation, while Figure 2b shows the Raman spectra after 6 hours incubation at 40 °C. The successful functionalization of the DNA strands can be identified by the observation of the ring breathing modes for Adenine and Cytosine in the Raman spectrum. Photoligation of the materials and subsequent characterization of the detection of DNA is part of our ongoing work in this area of research.

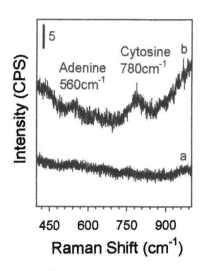

Figure 2. Detection of the DNA hybridization using Raman Spectroscopy. Before hybridization (A) and after 6 hours hybridization (B).

CONCLUSIONS

We have demonstrated a straightforward technique for the synthesis of monodispersed silver nanoparticles and their coating by a shell of gold of variable thickness that can be controlled by the amount of gold precursor used in the coating process. The use of acrylate as a dual reducing and encapsulating agent minimized the etching of metallic silver particles by the aqueous gold precursor. The use of these probes to detect DNA has been explored, the preliminary results are promising and further study of the system is part of our ongoing work.

ACKNOWLEDGMENTS

Derrick Mott gratefully acknowledges support by the Japan Society for the Promotion of Science (JSPS) fellowship. We thank Mr. Nobuaki Ito for assistance in the use of XPS instrumentation and Mr. Koichi Higashimine for assistance in the use of TEM instrumentation.

REFERENCES

1. Y-W. Cao, R. Jin and C. A. Mirkin, J. Am. Chem. Soc. 123, 7961 (2001).
2. L. Lu, A. Kobayashi, K. Tawa and Y. Ozaki, Chem. Mater. 18, 4894 (2006).
3. Z. S. Pillai and P. V. Kamat, J. Phys. Chem. B 108, 945 (2004).
4. Y. Cui, B. Ren, J-L. Yao, R-A. Gu and Z-Q. Tian, J. Phys. Chem. B 110, 4002 (2006).
5. V. K. Sharma, R. A. Yngard and Y. Lin, Adv. Colloid Interf. Sci. 145, 83 (2009).
6. N. T. B. Thuy, R. Yokogawa, Y. Yoshimura, K. Fujimoto, M. Koyano and S. Maenosono, Analyst 135, 595 (2010).

Mater. Res. Soc. Symp. Proc. Vol. 1272 © 2010 Materials Research Society   1272-PP06-20

Synthesis and Characterization of Carbon Nanotube-Nickel/Nickel Oxide Core/shell Nanoparticle Heterostructures Incorporated in Polyvinyl Alcohol Hydrogel

Wenwu Shi,[1] Kristy Crews,[2] Nitin Chopra[1]*
[1] Metallurgical and Materials Engineering, Center for Materials for Information Technology (MINT), the University of Alabama, Tuscaloosa, Al, 35487
[2] NSF-REU Fellow (2009), Chemistry, The University of West Alabama, Livingston, AL, 35470
* Corresponding author: E-mail: nchopra@eng.ua.edu , Tel: 205-348-4153, Fax: 205-348-2164

ABSTRACT

Carbon nanotube (CNT)-nickel/nickel oxide (Ni/NiO) core/shell nanoparticles (CNC) heterostructures were prepared in a unique single-step synthetic route by direct chemical precipitation of nanoparticles on CNT surface. Chemical vapor deposition (CVD)-grown CNTs (average diameter ~42.7±12.3 nm) allowed for direct nucleation and uniform coating of Ni/NiO core/shell nanoparticles (average diameter ~11.8±1.7 nm). The crystal structure, morphology, and phases in CNC heterostructures were studied using high resolution transmission electron microscopy (TEM), scanning electron microscopy (SEM), and X-ray photoelectron spectroscopy (XPS). Subsequently, the as-produced CNC heterostructures were incorporated into polyvinyl alcohol (PVA) hydrogel resulting in CNC heterostructure-PVA hydrogel with ~ 75% water absorbing capability. These novel hydrogels were also characterized by SEM and showed actuation under 0.2 T magnet. They are promising for smart analytical devices and platform.

INTRODUCTION

Unique properties of single-component nanostructures (e.g., carbon nanotubes (CNTs), nanowires, and nanoparticles) [1,2,3] are motivating researchers to develop multi-component and hybrid nanostructures [4-6]. Among these, CNT-nanoparticle heterostructures are promising for applications ranging from energy technologies to chemical and biological sensors [3,5,6]. Such heterostructures combine the effects of the individual components such as novel surface chemistry, stability in harsh environments, bioactivity, excellent mechanical strength, and optimum thermal, optical, magnetic, and electrical properties [3]. Interestingly, large curvature of CNTs also allows for the uniform packing of nanoparticles [5]. The approaches to synthesize CNT-nanoparticle heterostructures mostly rely on chemical functionalization methods [6-9]. However, the attachment of nanoparticles on CNTs without the use of covalent chemistry will be critical in obtaining non-contaminated heterostructures and would be desirable for biological applications. Towards this end, enhancing biocompatibility of these heterostructures is also important. Thus, incorporating such heterostructures in a biocompatible hydrogel will result in a hybrid, high strength, and multi-functional system. Hydrogels are porous and viscoelastic materials composed of cross-linked polymer that swell and shrink in presence and absence of an aqueous media [10-14]. For example, polyvinyl alcohol (PVA) hydrogel absorbs water and is biocompatible and non-toxic [14]. With only a few research efforts in encapsulating single-component nanostructures with hydrogels [15-17], assembling CNT-nanoparticle heterostructures in hydrogel remains elusive. This new area of research involves numerous challenges including development of an innovative synthetic process and acquiring fundamental knowledge concerning the physico-chemical characteristics of such hybrid materials.

Here, we report a unique synthetic approach for multi-component hydrogels comprised of CNT-nickel/nickel oxide (Ni/NiO) core/shell nanoparticles (CNC) heterostructures assembled inside PVA hydrogel (referred as 'CNC heterostructure-PVA hydrogel'). The CNC heterostructures were synthesized in a one-step approach by direct chemical precipitation of nanoparticles onto high surface area CNTs. Finally, CNC heterostructures were incorporated in PVA hydrogel and their water swelling characteristics was studied. There are numerous advantages of such heterostructures, including high surface-to-volume ratio, multi-functionality, and chemical selectivity. For example, robust CNTs are not magnetic but heterostructuring them with Ni/NiO core/shell nanoparticles resulted in magnetic CNC heterostructures. This unique single-step synthetic process also allowed for a uniform dispersion of magnetic nanoparticles onto CNT as substrates. Additionally, CNTs and Ni/NiO core/shell nanoparticles offer unique surface chemistries [5] critical for developing miniaturized and selective platforms. Further incorporation of CNC heterostructure into PVA hydrogel resulted in soft nanocomposite that could be actuated by applying magnetic field and demonstrated swelling/shrinking mechanism in presence/absence of water respectively. This unique nanocomposite hydrogels can be very useful in applications such as actuators, drug delivery systems, and separation media.

EXPERIMENTAL

Synthesis of multiwalled carbon nanotubes (CNTs): CNTs were grown by pyrolysis of xylene and ferrocene mixture under Ar/H$_2$ atmosphere in a tubular furnace [18]. A mixture of ferrocene (catalyst, 6.5 mol%, Sigma-Aldrich, St. Louis, MO) in xylene (hydrocarbon source, Fisher Scientific, Pittsburgh, PA) was preheated to 175-200 °C and CNTs were grown at 800 °C under the flow of argon and 10% hydrogen for 2 h. The as-produced multiwalled CNTs (or 'CNTs') were collected from the tube walls.

Synthesis of CNC heterostructures: Nickel nanoparticles were directly nucleated onto CNTs by thermal decomposition of nickel acetate (1.00 g, Sigma-Aldrich, St. Louis, MO) in 7 mL oleylamine (Sigma-Aldrich, St. Louis, MO) and 0.01 g CNTs under magnetic stirring. The mixture was heated in N$_2$ atmosphere at 90 – 95 °C for 40 min and followed by the addition of 1.50 g tri-n-octylphosphine oxide (TOPO, Alfa Aesar, Ward Hill, MA) and 1 mL trioctylphosphine (TOP, Sigma-Aldrich, St. Louis, MO). The temperature was slowly increased from 90 °C to 250 °C at a rate of ~10 °C/min and held at 250 °C for 30 min. The solution was cooled down naturally and rigorously washed using ethanol, hexane, and acetone (all purchased from Fisher Scientific, Pittsburgh, PA) and then centrifuged (Labnet Inc, Edison, NJ) at 5000-6000 rpm. The precipitate was dried overnight in a vacuum oven (VWR International, Suwanee, GA) at 80 °C to obtain black powder. The latter was exposed to air to form NiO shell around nickel nanoparticles and finally, resulting in CNC heterostructures.

Synthesis of CNC heterostructure-PVA hydrogel: 0.03 g dried CNC heterostructures were dispersed in 10 mL water by ultrasonicating and shaking for 120 min. Then, 0.8 g PVA (99+% hydrolyzed, M.W=89000~98000, Aldrich, St. Louis, MO) and 1 mL cross-linking agent (polyethylene glycol, PEG-600 solution, Alfa Aesar, Ward Hill, MA) were added to the CNC heterostructure solution. The mixture was heated in an oil bath at 95 °C for 1 hour, and then casted in a 4 inch petri dish. This sample was polymerized in a freeze (- 20 °C) –thaw (room temperature) process in a 24 h cycle. Control samples of pure PVA hydrogel without any heterostructures were also prepared in a similar method. The equilibrium water content for the hydrogels was calculated using:

$$\text{Equilibrium Water Content (EWC)} = 100\% \times \frac{W_{swollen\ hydrogel} - W_{shrunken\ hydrogel}}{W_{swollen\ hydrogel}}$$

Characterization methodology: Scanning Electronic Microscopy (SEM) images were obtained using FE-SEM JEOL-7000 at 20 kV for heterostructure and 10 kV for hydrogel. FEI Tecnai F-20 was used to collect Transmission Electronic Microscopy (TEM) images. X-ray photoelectron spectrum (XPS) was gathered by Kratos Axis 165 with Aluminum mono-gun at 160 pass energy.

RESULTS AND DISCUSSION

Critical to the purity of CNT-nanoparticle heterostructures is the elimination of chemical functionalization routes to link CNTs with nanoparticles. In this regard, we report a single-step synthetic route to directly nucleate Ni/NiO core/shell nanoparticles onto high curvature CNT surface resulting in CNC heterostructures. These CNC heterostructures were incorporated into PVA hydrogel (referred as CNC heterostructure-PVA hydrogel). Due to the ability of oleylamine to completely wet CNT surface, it was possible to nucleate nickel nanoparticles in a chemical precipitation reaction on the CNT surface. This chemical precipitation process involves the formation of nickel-oleylamine complex [19], using nickel acetate precursor, on CNTs and addition of TOPO and TOP led to the formation of CNT-nickel nanoparticle heterostructures. Air exposure of these heterostructures facilitated NiO shell formation on the nickel nanoparticles. Thus, resulting in CNC heterostructures as shown in figure 1.

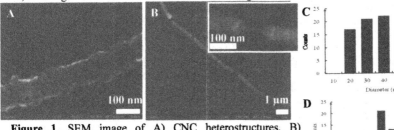

Figure 1. SEM image of A) CNC heterostructures, B) selectively coated Ni/NiO core/shell nanoparticles on CNT surface. Inset shows high resolution image with many aggregated small nanoparticles, C) and D) diameter distribution for CNTs and Ni/NiO core/shell nanoparticles, respectively.

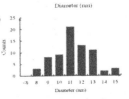

Figure 2. TEM image of A) CNC heterostructures. Inset shows high resolution image of heterostructures clearly indicating coated nanoparticles, B) Ni/NiO core/shell nanoparticles. Note: The NiO shell is polycrystalline and (111) interplaner spacing of nickel core is indicated.

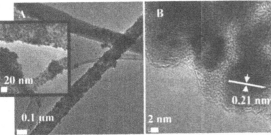

It was observed (Figure 1 B) that selective coating of Ni/NiO core/shell nanoparticles could be inherently achieved by disentangling CNC heterostructures from each other by rigorously washing and cleaning steps. It is proposed that overlapping CNTs prevented nanoparticle coatings on that particular region on CNTs and resulted in selective formation of CNC heterostructures. High resolution SEM and TEM studies (Figure 1 (inset) and 2) showed that the nanoparticles were uniformly distributed and tightly packed onto CNT surface. Average CNT diameter was estimated to be ~ 42.7±12.3 nm and that of Ni/NiO core/shell nanoparticles to be ~ 11.8±1.7 nm (Figure 1C and D). The NiO shell thickness of ~ 2.1±0.4 nm was observed.

In order to determine the surface bond state, structural defects, and chemical composition of the CNC heterostructures, detailed XPS studies were done (Figure 3). All spectra were fitted by XPS peak 4.0 software [20]. For as-produced CNTs, XPS showed (C 1s, Figure 3 A) graphitic structure of CNTs (284.48 eV), defects created by virtue of CVD synthesis method (285.43 eV), and weak carbon-oxygen covalent interactions (288.78eV, 286.78 eV), and $\pi - \pi^*$ transition (291.2 eV) [21]. For O 1s (Figure 3 B) peaks in CNTs, two peaks, 531.92 eV (–OH) and 533.2 eV (-COOH) were obtained [21]. As-produced CNTs without acid treatment or oxidation resulted in low intensity and signal-noise ratio for the O 1s peaks. As for the CNC heterostructure, strong graphitic CNT peak (284.94 eV) but slightly shifted from those in figure 3 A is observed. This shift can be attributed to the interfaces between CNT and Ni/NiO core/shell nanoparticles. In addition, weak C-H and C-O peaks were also observed (288.19 eV, Figure 3 C). Nevertheless, contributions from remaining surfactants (TOPO and TOP) used for nanoparticle synthesis should not be over-ruled. Surfactants were not fully eliminated and thus, P 2p peaks (Figure 3 F, P $2p_{3/2}$:130.08 eV and $2p_{1/2}$:131.35 eV) were observed [22]. Another peak (133.56 eV) was corresponding to P-O from TOPO was observed [23]. For CNC heterostructures, O 1s spectrum (Figure 3 D) consisted of two peaks, 533.64 eV and 533.10 eV corresponding to Ni-O and –OH bond on the surface of NiO, respectively [24,25]. For Ni 2p (Figure 3 E) indicated $2p_{3/2}$ (853 eV), $2p_{1/2}$ (870.37 eV) for metallic Ni^0 (Ni core), $2p_{3/2}$ (855.8 eV), $2p_{1/2}$ (874.03 eV) for Ni^{2+} (NiO shell), and two other satellite peaks (861.36 eV and 879.52 eV) [24,25]. Moreover, N 1s spectra (Figure 3 G) for CNC heterostructure showed no obvious peaks, which confirmed absence of oleylamine. Such multi-functional heterostructures are magnetic as shown by actuating them under a 0.2 T magnet (Figure 4D and E) and hold great promise for biological and mechanical devices. In this regard, CNC heterostructures were incorporated into a biocompatible PVA hydrogel. Figure 4 A, B and F shows shrunken hybrid hydrogel with uniformly dispersed CNC heterostructures. Prepared hydrogels have good swelling property at room temperature in DI water. Equilibrium water content (figure 4C) of ~ 75% was achieved for CNC heterostructure-PVA hydrogel. This was 6-33% lower than pure PVA hydrogels due to the presence of CNC heterostructures in the hybrid hydrogels that resulted in reduced pore fraction.

CONCLUSIONS

CNC heterostructures were fabricated in a single-step synthetic approach by direct nucleation of Ni/NiO core/shell nanoparticles on the surface of CVD-grown CNTs. Uniform dispersion and ideal packing of core/shell nanoparticles was achieved on the CNTs. These heterostructures were then thoroughly characterized for their crystal structure, morphology, phases, and surface functional groups using SEM, TEM, and XPS. Finally, CNC heterostructure were uniformly incorporated into a PVA hydrogel resulting in novel and magnetic CNC heterostructure-PVA hydrogel with ~ 75% of water absorbing capacity. These hybrid hydrogels are promising for multi-functional chemical and biological sensors, protein separation, drug delivery, and smart analytical platform.

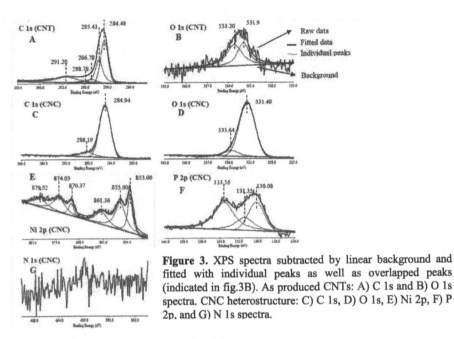

Figure 3. XPS spectra subtracted by linear background and fitted with individual peaks as well as overlapped peaks (indicated in fig.3B). As produced CNTs: A) C 1s and B) O 1s spectra. CNC heterostructure: C) C 1s, D) O 1s, E) Ni 2p, F) P 2p, and G) N 1s spectra.

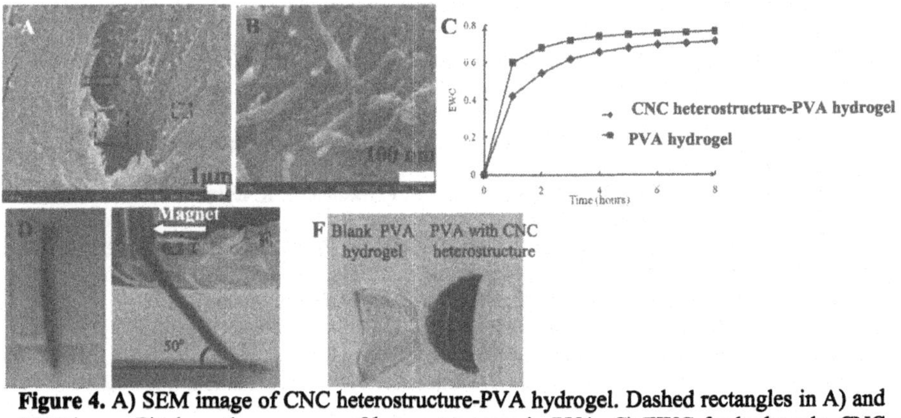

Figure 4. A) SEM image of CNC heterostructure-PVA hydrogel. Dashed rectangles in A) and SEM image B) show the presence of heterostructures in PVA. C) EWC for hydrogels. CNC heterostructure-PVA hydrogel in swollen state C) when no magnet is near it and D) when 0.2 T magnet is brought near, the hydrogel actuates by an angle of 50°. Digital images showing E) blank PVA hydrogel and F) CNC heterostructures-PVA hydrogel with uniform dark color.

ACKNOWLEDGEMENTS

The authors thank the University of Alabama (UA) for the start-up funds, NSF (ECCS-PCAN 0925445), and Research Grant Committee (RGC) Award for supporting this work. K. C. thanks NSF-REU program. The authors also thank the Central Analytical Facility (CAF) for access to the electron microscopy equipment (NSF funded), the MINT Center for providing infrastructure support such as clean room facility and various equipments, CAF staff, Junchi Wu for providing training on SEM and XPS. Thanks to Dr. S. Kapoor for editing this manuscript.

REFERENCES

1. C. J. Murphy, T. K. Sau, A. M. Gole, C. J. Orendorff, J. Gao, L. Gou, S. E. Hunyadi and T. Li, *J. Phys. Chem. B* **109**,13857 (2005).
2. J. Hu, T. W. Odom and C. M. Lieber, *Acc. Chem. Res.* **32**, 435 (1999).
3. M. Meyyappan, "Carbon nanotubes: Science and applications", CRC Press LLC (2005, Boca Raton, FL).
4. L. J. Lauhon, M. S. Gudiksen and C. M. Lieber, *Phil. Trans. R. Soc. Lond. A* **362**, 1247 (2004).
5. N. Chopra, L. Claypoole and L. G. Bachas, *NSTI Nanotech 2009 Proc.* **1**, 187 (2009).
6. N. Chopra, M. Majumder and B. J. Hinds, *Adv. Funct. Mater.* **15**, 858 (2005).
7. X. Peng, J. Chen, J. A. Misewich and S. S. Wong, *Chem. Soc. Rev.* **38**, 1076 (2009).
8. B. M. Quinn, C. Dekker and S. G. Lemay, *J. Am. Chem. Soc.* **127**, 6146 (2005).
9. V. Tzitzios, V. Georgakilas, E. Oikonomou, M. Karakassides and D. Petridis, *Carbon*, **44**, 848 (2006).
10. R. Zhang, A. Bowyer, R. Eisenthal and J. Hubble, *Biotech. Bioeng.* **97**, 976 (2006).
11. Y. Qiu and K. Park, *Adv. Drug Del. Rev.* **53**, 321 (2001).
12. N. A. Peppas, J. Z. Hilt, A. Khademhosseini and R. Langer, *Adv. Mater.* **18**, 1345 (2006).
13. R. Langer and D. A. Tirrell, *Nature* **428**, 487 (2004).
14. N. A. Peppas and S. R. Stauffer, *J. Control. Rel.* **16**, 305 (1991).
15. V. Pardo-Yissar, R. Gabai, A. N. Shipway, T. Bourenko and I. Willner. Gold, *Adv. Mater.* **13**, 1320 (2001).
16. V. Kozlovskaya, E. Kharlampieva, B. P. Khanal, P. Manna, E. R. Zubarev and V. V. Tsukruk, *Chem. Mater.* **20**, 7474 (2008).
17. J. Shi, Z.-X. Guo, B. Zhan, H. Luo, Y.Li and D. Zhu, *J. Phys. Chem. B*, **109**, 14789 (2005).
18. N. Chopra, P. D. Kichambare, R. Andrews and B.J. Hinds, *Nano Lett.* **2**, 1177 (2002).
19. N. Chopra, L. Claypoole and L. G. Bachas, *J. Nanopart. Res.* In press (2010).
20. R.W.M. Kwok, XPS Peak Fitting Program for WIN95/98 XPSPEAK Version 4.1, Department of Chemistry, The Chinese University of Hong Kong.
21. V. Datsyuk, M. Kalyva, K. Papagelis, J. Parthenios, D. Tasis, A. Siokou, I. Kallitsis and C. Galiotis, *Carbon* **46**, 833 (2008).
22. J. F. Moulder, W. F, Stickle, P. E. Sobol and K. D Bomben, Handbook of X-ray Photoelectron Spectroscopy, 1995, Physical Electronics, Inc.
23. A. A. Guzelian, J. E. B. Katari, A. V. Kadavanich, U. Banin, K. Hamad, E. Juban, A. P. Alivisatos, R. H. Wolters, C. C. Arnold and J. R. Heath, *J. Phys. Chem.* **100**, 7212 (1996).
24. M. Salavati-Niasari, F. Mohandes, F. Davar, M. Mazaheri, M. Monemzadeh and N. Yavarinia, *Inorg. Chim. Acta* **362**, 3691 (2009).
25. L. A. Garcia-Cerda, L. E. Romo-Mendoza and M.A. Quevedo-Lopez, *J Mater. Sci.* **44**, 4553 (2009).

Mater. Res. Soc. Symp. Proc. Vol. 1272 © 2010 Materials Research Society 1272-PP07-02

Enhancing cell culture in magnetic vesicle gels

Felicity Leng[1,2], Julie E. Gough[2] and Simon J. Webb[1]
[1]School of Chemistry and Manchester Interdisciplinary Biocentre, University of Manchester, 131 Princess St, Manchester M1 7DN, U.K.
[2]School of Materials, University of Manchester, Grosvenor St, Manchester, M1 7HS, U.K.

ABSTRACT

Several different hydrogel compositions have been incorporated into magnetic vesicle gels and the resulting "smart" biomaterials assessed as cell culture scaffolds. The compatibility of these hydrogels with the "smart" component of these biomaterials, thermally sensitive vesicles (TSVs) crosslinked by magnetic nanoparticles, was assessed by the leakage of fluorescent 5/6-carboxyfluorescein from the TSVs under cell culture conditions. Subsequently the ability of the hydrogels to support 3T3 fibroblast and chondrocyte viability was assessed. These studies revealed that alginate-based gels were the most compatible with both the TSVs and the cultured cells, with an alginate:fibronectin mix proving to be the most versatile. Nonetheless these studies also suggest that TSV composition needs to be modified to improve the performance of these "smart" cell culture scaffolds in future applications.

INTRODUCTION

Creating "smart" biomaterials that are able to replicate the complex structure of tissue and chemically communicate with cells cultured within them has thus far proved to be a challenging task. To this end we recently developed magnetic vesicle gels, a new type of biomaterial composed of adhesive lipid-doped thermally sensitive vesicles (TSVs) crosslinked with magnetic nanoparticles. The resulting magnetic nanoparticle-vesicle assemblies are then embedded in a biocompatible hydrogel to give a magnetic vesicle gel.[1] The magnetic functionality in these materials allows non-invasive magnetospatial control of vesicle-nanoparticle assemblies[2] in the gel and facilitates magnetic release of the vesicle contents into the surrounding volume. The application of an alternating magnetic field (AMF) releases bioactive compounds stored in the vesicles, which then diffuse through the gel and trigger cellular responses.

A key part in the design of these materials is the hydrogel scaffold that immobilizes the magnetic nanoparticle-vesicle assemblies and provides the local environment that supports cell growth. Previously we used a calcium alginate scaffold, as we found this material held the assemblies in place without the gel fibrils disrupting the TSV membranes.[1] Alginate gels are also easily formed and manipulated at physiological temperatures, which was hoped to allow these magnetic vesicle gels to be applied to biological systems. Nonetheless, calcium alginate is a poor scaffold for the proliferation of several types of cell. To improve the versatility of these vesicle gels for cell culture applications, several other gel scaffolds were tested for compatibility with TSVs and several cell lines.

EXPERIMENTAL DETAILS

N-(Biotinoyl)dopamine was synthesized by a modification of literature procedures.[3] Magnetic nanoparticles (MNPs) were formed as previously detailed[2] from iron(II) chloride tetrahydrate and iron (III) chloride hexahydrate. These iron salts were dissolved in deoxygenated water then added dropwise to NaOH solution (1 M) under nitrogen with vigorous stirring over 30 minutes. The particles were then magnetically sedimented using a N48 5350 G neodymium iron boron magnet and washed to remove any nonmagnetic material. The magnetic nanoparticles (12.5 mg) were suspended in methanol (3 mL) under nitrogen, then mixed with N-(biotinoyl)dopamine (3.5 mg, 9.2×10^{-6} mol) and sonicated for 4 hours. The coated particles were magnetically sedimented using an NdBFe magnet and washed with methanol repeatedly, using a magnet to sediment the particles between washes, to give coated MNPs as a fine brown powder.

Thermally sensitive vesicles (TSVs) were composed of 4:1 mol/mol dipalmitoyl phosphatidylcholine (DPPC) and dimyristoyl phosphatidylcholine (DMPC), with 1 % mol/mol triethylammonium (N-(biotinoyl)-1,2-dihexadecanoyl-sn-glycero-3-phosphoethanolamine (Bt-DHPE, from Sigma) added if the TSVs were to be crosslinked with magnetic nanoparticles. The lipids (DMPC, 1.27 mg; DPPC, 12.55mg; Bt-DHPE, 200μL from 1 mM stock solution in CHCl$_3$ if required) were dissolved in chloroform (1 mL) and the solvent removed under reduced pressure to give a lipid thin film on the inside of the round-bottomed flask. The lipid film was hydrated in 0.05M 5/6-carboxyfluorescein (5/6-CF) in MOPS buffer at pH 7.4 (1 mL) then extruded through a single 800 nm polycarbonate membrane (19 ×) at > 40 °C to give ~800 nm diameter biotin-tagged TSVs.

The magnetic nanoparticle-vesicle assemblies were formed immediately prior to use by mixing biotin coated nanoparticles and the biotin-tagged TSVs with avidin in a 2:1 biotin:avidin ratio, and allowed to aggregate for 60 min. Magnetic sedimentation was used to remove unencapsulated 5/6-CF and non-aggregated vesicles. The vesicle suspensions were placed upon a 5 kG magnet until a compact vesicle plug had formed at the bottom of the vial and the supernatant solution was visually free of turbidity. As much of the supernatant was removed as possible without disturbing the TSV plug (typically 60% of the volume), and replaced with an equal volume of the appropriate buffer solution. Brief vortex mixing regenerated the vesicle suspension. This procedure was repeated at least 6 times and until the concentration of unencapsulated 5/6-CF was < 0.1% of the initial concentration.

The cells were seeded in gel scaffolds at a cell count of 5 cells/μL for 3T3 fibroblasts and 10 cells/μL for chondrocytes; 3T3 fibroblasts proliferate much faster than chondrocytes and were therefore seeded at a lower cell count, generally 20,000 vs. 40,000. The gel matrix was then added to the cell suspension in thin films of gel (0.500 mL per well in 24 well plates and 0.800 mL per well in 12 well plate). The gel scaffolds were then suspended in fresh DMEM media to allow transfer of cell nutrients and waste. The metabolic activity and DNA count of the cell lines were measured to study the reaction of the cells to their surroundings.

Alginate gels were formed by stirring alginic acid (2 % wt/vol) in PBS for 60 minutes at room temperature. The gel was formed by infusion of CaCl$_2$ (0.1 M) through a 800 nm polycarbonate membrane. If required, fibronectin solution (1 mg/mL) was added at 1 % vol/vol. The gels were then sterilized by autoclaving at 120°C for 30 minutes. Chitosan was added to well plate, then cured using PBS whilst adding cell solution into the gel. Chitosan:PEI gels were formed by suspending cells in PEI solution (30 % wt/vol in PBS, pH 11) and then adding this solution to chitosan (2 % wt/vol in aqueous acetic acid, pH 3); the gel formed upon mixing. Cells were suspended in gelatin solution (1 % wt/vol), then cooled to room temperature for gelation.

DISCUSSION

Testing new hydrogels for compatibility with DPPC:DMPC TSVs

In addition to calcium alginate, four other types of gel that have been used for cell culture were tested for compatibility with TSVs.[4,5,6,7] Mixing chitosan and polyethyleneimine (PEI) in a 50:50 ratio under physiological conditions will lead to stable gels that have been shown to support cell proliferation. This procedure does not require external stimuli such as heat, sonication or pH changes, which would disrupt the membranes of the TSVs in the gel. Similarly chitosan itself, although it requires a pH change to gel,[8] allows sufficient time to add the TSVs to the mixture after buffer addition (phosphate buffer, pH 7.4) before the hydrogel gel forms. Gelatin, a widely used hydrogel scaffold, was also tested. Although heating is required for gelation, after heating the gelatin remains fluid at room temperature for 5 minutes, allowing TSVs to be added at 25 °C before full gelation occurred. Finally, as alginate has already been established as a good scaffold for magnetic nanoparticle-vesicle assemblies, this material was modified to increase cell attachment and proliferation by doping the alginate matrix with fibronectin. Fibronectin, a high molecular weight glycoprotein, is found in the extracellular matrix and can significantly enhance cell adhesion, also playing a role in cell differentiation and migration.[4]

To test for vesicle membrane disruption during the gelation of these mixtures, TSVs with encapsulated carboxyfluorescein were mixed with the hydrogelator in buffer. The solution was allowed to gel and the change in fluorescence was monitored over time under cell culture conditions (37 °C in a 5 % CO_2 atmosphere), with an increase in fluorescence indicating that the gelation process had disrupted the vesicle membrane.

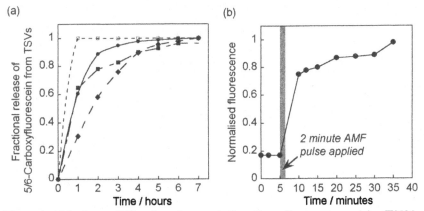

Figure 1: (a) Leakage of 5/6-carboxyfluorescein from thermally sensitive vesicles (TSVs) embedded in different gel matrices at 37 °C (♦) Alginate:fibronectin (○) Chitosan (●) Chitosan:PEI (×) Gelatin (■) Alginate. (b) Magnetic release of 5/6-carboxyfluorescein encapsulated in magnetic nanoparticle-vesicle assemblies (in suspension, MOPs buffer pH 7.4), after exposure to a 2 minute pulse of 392 kHz alternating magnetic field (AMF) at 5 minutes.

At 37 °C there was significant leakage of the TSV contents in all gel mixtures over a 6 hour period (Fig. 1 (a)), which can be ascribed to the proximity of the T_m of this TSV composition to the temperature required for cell culture. This is despite the 1-2 degree increase in T_m anticipated from previous studies of these 4:1 DPPC:DMPC vesicles in alginate gels.[1] Chitosan and gelatin were the least suitable gel matrices as gelation of these mixtures caused 100% release of the encapsulated dye by the first time point (1 hour). The high rate of leakage from the TSVs in chitosan may be due to incomplete neutralization of the gel, with the low pH causing extensive membrane disruption. The gelatin gels also showed poor stability under these conditions. The other gel matrices showed little difference in the rate of dye release. Interestingly the undoped alginate gel showed slightly increased rate of leakage compared to alginate doped with 1% fibronectin, which may indicate some bilayer-protein interactions.

Clearly the conditions used to form chitosan and gelatin gels are incompatible with the inclusion of TSVs in these gels, but this 4:1 DPPC:DMPC TSV composition also appears to have a T_m that is too low to be useful for cell culture in magnetic vesicle gels. Despite the relatively rapid rate of contents release from this TSV composition under magnetic stimulation (release of 5/6-CF is shown in Fig 1 (b)), the response of the cells to the released stimuli is likely to take hours to days, making the observed background rate of release too high in all hydrogels.

Testing new hydrogels for compatibility with 3T3 fibroblasts and chondrocytes

The three hydrogel scaffolds that were shown to give the slowest release of encapsulated contents at 37 °C (alginate, alginate-fibronectin, chitosan:PEI) were tested for their ability to support the proliferation of two different cell lines; 3T3 fibroblasts and chondrocytes.

The 3T3 cell line is a robust cell line of Swiss mouse fibroblasts with a rapid proliferation rate, making this an excellent cell line to trial our methodology. Furthermore the effect of the gel composition on cell proliferation could clearly be analyzed using the live/dead assay.

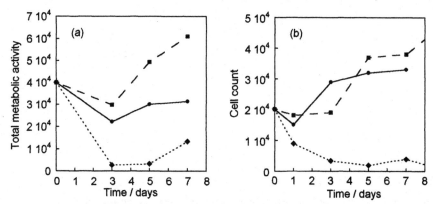

Figure 2: (a) Metabolic activity of 3T3 fibroblasts on gel scaffolds as determined by the alamar blue assay; (b) Proliferation of 3T3 fibroblasts on gel scaffolds as determined by the DNA assay; (■) Alginate/fibronectin (●) Alginate (♦) Chitosan/PEI.

224

Cell counts on the gel samples show that doping the alginate gel with fibronectin significantly increases metabolic activity and the rate of cell proliferation, while the undoped alginate scaffold supports cell proliferation much better than chitosan-PEI. All these gel scaffolds were however much less effective than tissue culture plastic at supporting cell growth, with even the alginate-fibronectin mixture giving less than 10% of the cell count of this control. The chitosan:PEI gels also presented several experimental problems, as they were difficult to seed, had poorly reproducible gel stiffness and interfered with assays using pH sensitive dyes like alamar blue. The morphology of the cells studied by live/dead assays shows that when fibronectin is present the cells attach to the surrounding scaffold and spread, while in alginate only cells remain rounded with no attachment points.

When chondrocytes were cultured on these same three scaffolds (alginate, alginate:fibronectin and chitosan:PEI), the alamar blue metabolic activity assay showed that the activity of these cells generally increased over time as the cells proliferated (Figure 3 (a)). Little difference in chondrocyte activity was observed between the undoped alginate and the fibronectin doped alginate scaffold, possibly because chondrocytes can to some extent maintain themselves in a rounded form and the presence of RGD attachment points in the matrix is not a crucial as for fibroblasts.[9] In addition, the external addition of calcium(II) to either chondrocytes or fibroblasts revealed that chondrocytes appeared to have a higher tolerance for calcium(II) in the surrounding matrix. In comparison the DNA assay of cell number shows that the undoped alginate scaffold gave the best cell proliferation (Figure 3 (b)), followed by the alginate:fibronectin mixture. Chitosan:PEI scaffolds did initially support faster proliferation of chondocytes than the alginate based gels, but this sharply declined after a few media changes. In comparison few cells survived exposure to pure chitosan gel; separate tests on pure chitosan scaffolds showed no cell proliferation until 10 days had passed.

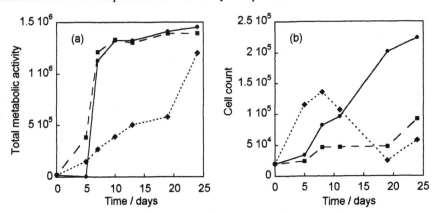

Figure 3: (a) Metabolic activity of chondrocytes on gel scaffolds as determined by the alamar blue assay; (b) Proliferation of chondrocytes on gel scaffolds as determined by DNA assay; (■) Alginate/fibronectin (●) Alginate (♦) Chitosan/PEI.

CONCLUSIONS

Of the five hydrogels tested for compatibility with thermally sensitive vesicles (TSVs), 3T3 fibroblasts and chondrocytes, alginate-based gels showed the greatest versatility for the development of new types of magnetic vesicle gels. Both gelatin and chitosan gels were found to be incompatible with TSVs and caused extensive membrane disruption, while further tests on chitosan gels also showed that they did not support cell growth. Mixing chitosan with PEI produced gels that were more cell-compatible, but the materials properties of these mixed gels were inconsistent and the gels proved to be poorer supports for cell growth than alginate gels. The chief drawback of undoped calcium alginate gels was the lack of cell attachment points in the matrix. This was a less significant problem for chrondrocytes than for fibroblasts, but the addition of fibronectin at 1 % gave significant improvements in 3T3 fibroblast proliferation without significantly changing TSV compatibility. Nonetheless studies on dye leakage from TSVs in alginate gels at 37 °C revealed that the 4:1 DPPC:DMPC composition in the magnetic vesicle gels that were developed previously would be unsuitable for cell culture. Future development of these "smart" biomaterials will require TSV compositions with higher melting temperatures. These high T_m TSVs will be mixed with calcium alginate gels and cell viability in the resulting magnetic vesicle gels assessed under physiological conditions.

ACKNOWLEDGMENTS

FL would like to thank the BBSRC DTG programme for funding.

REFERENCES

1. R.J. Mart, K.P. Liem and S.J. Webb, *Chem. Commun.*, 2287, (2009).
2. K.P. Liem, R.J. Mart and S.J. Webb, *J. Am. Chem. Soc*, **129**, 12080, (2007).
3. H. Gu, Z. Yang, J. Gao, C. K. Chang, and B. Xu, *J. Am. Chem. Soc.* **127**, 34, (2005).
4. A. Mosahebi, M Wiberg and G. Terenghi, *Tissue Eng*, **9**, 2, 209 (2003).
5. C.L. Huang, J.D. Liao, C.F. Yang, C.W. Chang, M.S. Ju and C.C.K. Lin, *Thin Solid Films* **517**, 5386 (2009).
6. F. Khan, R. Tore and M. Bradley, *Angew. Chem. Int. Ed.*, **47**, 1, (2008).
7. G.V.N. Rathna, *J. Mater. Sci. - Mater. M.*, **19**, 2351 (2008).
8. J.K. Francis Suh and H.W.T. Matthew, *Biomaterials*, **21**, 2589, (2000).
9. Y.J. Lin, C.N. Yen, Y.C. Hu, Y.C. Wu, C.J. Liao and I.M. Chu., *J. Biomed. Mater. Res. A*, **88A**, 23, (2009).

Mater. Res. Soc. Symp. Proc. Vol. 1272 © 2010 Materials Research Society   1272-PP07-04

PEG-based bioactive hydrogels crosslinked via phosphopantetheinyl transferase

Katarzyna A. Mosiewicz[1], Kai Johnsson[2] and Matthias P. Lutolf[1]
[1]Laboratory of Stem Cells Bioengineering, Institute of Bioengineering, Ecole Polytechnique Fédérale de Lausanne, CH-1015 Lausanne, Switzerland
[2]Laboratory of Protein Engineering, Institute of Bioengineering, Ecole Polytechnique Fédérale de Lausanne, CH-1015 Lausanne, Switzerland

ABSTRACT

State-of-the-art tissue engineering strategies increasingly rely on the performance of bioactive hydrogels formed via cell-friendly crosslinking reactions. Enzymatic reactions possess ideal characteristics for such applications, but they are currently still underexplored in biomaterials design. Here we report the development of hybrid bioactive hydrogels formed via a posttranslational modification reaction using phosphopantetheinyl transferase (PPTase). PPTase was shown to catalyze the covalent crosslinking of CoenzymeA-functionalized poly(ethylene glycol) (PEG) multiarm macromers and recombinantly produced acyl carrier protein (ACP) dimers. Crosslinking kinetics and physicochemical properties of PPTase hydrogels were characterized. Proof-of-principle experiments demonstrate the successful covalent bio-functionalization of gels with a CoA-derivatized cell adhesion peptide. Polymerization of gels in the presence of primary mammalian cells was shown to result in no loss in cell viability compared to a well established, chemically crosslinked gel system.

INTRODUCTION

The tissue-specific milieu that harbors cells in our tissues is critical for their function. A complex mixture of extracellular signaling cues delivered by neighboring cells and the extracellular matrix (ECM) regulates key cell functions including cell adhesion, migration, proliferation and differentiation. The rapid increase in the molecular understanding of the regulatory role of this cell–ECM crosstalk in recent years has opened the door for the design of novel families of 'smart' biomaterials, which allow recapitulating key functions of natural ECM to manipulate cell behavior in a well-controllable manner. Owing to their structural and physicochemical similarities to natural ECM, synthetic hydrogels have shown to be particularly promising candidates as scaffolds for cell biology and tissue engineering.

An important criterion for the performance of synthetic hydrogels applied in biological contexts is a suitable chemical strategy to enable the self-selective gel cross-linking as well as the tethering of desired biomolecules to the gel under physiological conditions. Such crosslinking reactions allow forming gels *in situ*, that is, in the presence of biomolecules or cells present in tissues.

Apart from a few selective covalent reactions including Michael-type addition reactions[1], click chemistry[2] or native chemical ligation[3], enzymatic reactions, which are specific by their very nature, are ideal targets to catalyze gel formation. Surprisingly, only relatively few enzymatic reactions including transglutaminases and oxidoreductases [4-8] have been utilized for hydrogel engineering. A greater focus on these reactions could possibly result in advanced synthetic biomaterials with greater performance in clinical tissue engineering or (stem) cell biology.

Here we focus on phosphopantetheinyl transferase (PPTase) as gel-forming enzyme. PPTase catalyzes the covalent transfer of the 4-phosphopantetheine moiety of coenzyme A (CoA) to the serine residue at the active site of carrier proteins (CPs), a key posttranslational modification in the biosynthesis of fatty acids, polyketides and nonribosomal peptides [9]. The small size of CPs (<85 residues), the availability of CoA, and the ease of PPTase production are attractive features to utilize this enzyme in biomaterials engineering. Furthermore, short CP-mimicking peptides have been reported and successfully applied in multicolor live-cell imaging or functional protein microarrays[10, 11].

We have synthesized poly(ethylene glycol) (PEG)-based precursors suitable for PPTase-catalyzed gel crosslinking. Two gel building blocks were used, namely CoA-functionalized multiarm PEG and a recombinant apo-acyl carrier protein (ACP) dimer containing two active sites. As PPTase we chose surfactin synthetase Sfp of *Bacillus subtilis* orgin[9], which we produced in *E. coli*. The activity of related carrier proteins for Sfp ($k_{cat}/K_m = 0.2$-2.5 $\mu M^{-1}min^{-1}$[10, 12]) is comparable to substrates of the transglutaminase FXIIIa ($k_{cat}/K_m = 0.56$ $\mu M^{-1}min^{-1}$) which we have previously used for PEG gel crosslinking[8]. In the following we characterize PPTase gel crosslinking kinetics and physicochemical properties, as well as the culture of cells on (2D) or within (3D) gels functionalized with the integrin-binding cell adhesion peptide RGDS.

EXPERIMENTAL

Branched PEG (8-arm, 40 kDa, NOF corporation) or monomethoxy-PEG (5kDa) was functionalized with vinylsulfone end-groups as descried[13]. Coenzyme A disodium salt (Sigma) was conjugated to PEG via Michael-type addition of thiol-containing phosphopantetheine group of coenzyme A (1.2: 1 molar ratio of CoA over vinylsulfone, 0.3 M triethanolamine, pH 8.0) for 2 hours at room temperature (Fig.1a). Excess of Coenzyme A was removed by dialysis (Spectra/Por 6 MWCO 8 kDa) at 4 °C against 50mM ammonium acetate buffer (3 days) and against water (2 days). The resulting product was tested for purity using Size Exclusion Chromatography (Biosuite UHR 250 4.6x300mm column). No free CoA ($t_R = 11$ min) was detected. PEG-CoA elutes at the different retention time ($t_R = 7.6$ min) and is detectable by UV (260 nm). Efficiency of coupling was determined by ^1H NMR (400 MHz, D_2O) (Fig. 1c). Upon coupling, characteristic VS-peaks (Fig. 1c - left) δ 6.3 ppm (d, 1H, $J = 10$ Hz, -CH=$CH_{2\,cis}$), 6.4 ppm (d, 1H, $J = 16.8$ Hz, CH=$CH_{2\,trans}$) and 6.8 ppm (dd, 1H, $J = 10$ Hz, -SO_2CH=) disappeared, indicating that all vinylsulfone groups had reacted, while CoA-derived hydrogen peaks occurred (Fig. 1c - right).

To form hydrogels from 8-PEG-CoA via PPTase catalysis, the bi-functional ACP-linker protein ACP_ACP (Fig. 2a, b) was genetically engineered to contain a 6xHis-tag (in yellow), followed by two ACPs (blue), separated by two amino acids. A gene encoding ACP_ACP was inserted into pET15B and the resulting plasmid was transformed by electroporation into *E.coli* strain BL21(DE3). Protein expression was induced by the addition of 1 mM isopropyl-b-D-thiogalactosidase (IPTG). The bacteria were grown for 3 h at 180 rpm at 30 °C and then harvested by centrifugation and lysed by sonication. Purification of ACP_ACP was conducted using Ni-NTA (Qiagen) (Fig. 2c). Each step of expression was analyzed by SDS-PAGE (Fig. 2c). Eluted protein was dialyzed against 10 mM HEPES buffer (pH 7.6) at 4 °C. The

228

concentration of 6xHis-ACP-ACP (MW 19.5 kDa) was determined by its extinction coefficient (ε_{280nm} = 2980 $[M^{-1}cm^{-1}]$, ExPASyProtParam).

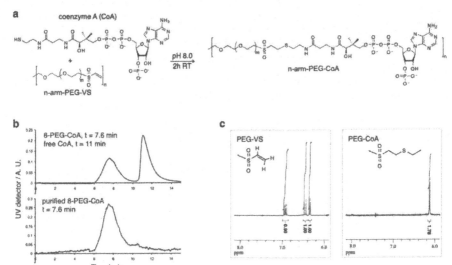

Figure 1. a, Scheme of Coenzyme A conjugation to monomethoxy-PEG-VS (5 kDa) or 8-arm PEG-VS macromer (40 kDa) via Michael-type addition. **b,** Chromatogram of PEG-CoA and free Coenzyme A eluted at different retention times using Size Exclusion Chromatography with UV detection at 260 nm. **c,** ^{1}H NMR of 8-arm PEG-VS macromer (left) and its Coenzyme A-modified version (right) formed via Michael-type addition. Parts of the figure adapted with permission from Reference [14]. Copyright 2010 American Chemical Society.

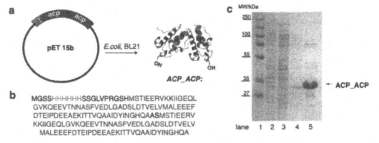

Figure 2. a, Construct of DNA-vector encoding ACP-dimer (ACP_ACP) containing a N-terminal His-tag for purification. **b,** Amino-acid sequence of ACP_ACP in a single letter code. **c,** SDS-PAGE analysis of protein expression and purification; lane 1 – marker (PageRuler), 2 – insoluble fraction of lysate, 3 – soluble fraction, 4 – flow through Ni-NTA resin, 5 – eluted protein.

RESULTS AND DISCUSSION

We first explored PPTase-catalyzed conjugation of ACP by CoA-modified PEGs in solution. The fluorescein (Fl)-labeled fusion protein ACP-SNAP was chosen for detecting the crosslinking reaction, as SNAP can be readily tagged with a fluorescent marker[15]. Upon mixing of ACP-SNAP-Fl and CoA-modified PEGs and in the presence of PPTase, a marked shift in molecular weight was detected (Fig. 3a), demonstrating successful bio-conjugation of both polymers.

Next, the dual functionality of the ACP_ACP linker protein was confirmed by its reaction with CoA-fluorescein (CoA-Fl) and 1-PEG-CoA (Fig. 3b).

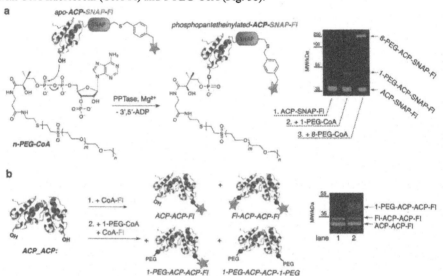

Figure 3. a, linear (n=1; 1-PEG-CoA; 5 kDa) and branched PEG-CoA (n=8; 8-PEG-CoA; 40 kDa) reacted with a fluorescent ACP fusion protein (ACP-SNAP-Fl, 34 kDa) by PPTase-catalyzed transfer of phosphopantethein prostetic group of CoA to the serine residue of apo-ACP; SDS-PAGE and laser-based in-gel fluorescence scanning show corresponding molecular weight shifts. **b**, ACP_ACP reacted with CoA-Fl (~1 kDa) (lane 1) or both CoA-Fl and 1-PEG-CoA (lane 2). Parts of the figure adapted with permission from Reference [14]. Copyright 2010 American Chemical Society.

Hybrid polymer gels were successfully created by mixing two stoichiometrically balanced liquid precursors containing ACP_ACP and 8-PEG-CoA (Fig. 2). At 37 °C and neutral pH in Hepes buffer containing 10 mM magnesium chloride and 1.0 mM of PPTase, a 5% (w/v) precursor solution gelled within approximately 15 min (Fig. 4b). All negative control conditions failed in forming hydrogels (Table 1).

Table 1. Specificity of PPTase-system shown in series of controls.

CONDITION:	1	2	3	4	5	6	7
ACP-ACP	-	+	+	+	+	+	+
PEG-CoA	+	-	+	+	2x excess	+	+
PPTase	+	+	-	+	+	Lyophil.	+
MgCl₂	+	+	+	-	+	+	+
OUTCOME:	No gel	No gel	No gel	No gel	No gel	No gel	GEL

We measured a striking decrease of the gel point as a function of increasing PPTase concentration (Fig. 4a). Small strain oscillatory shear rheometry on fully crosslinked hydrogels revealed a frequency dependence of G' and G" that is typical of elastic, covalently cross-linked polymer gels, with a storage modulus (G' = 2.3 kPa) that was higher than the loss modulus (G" = 0.6 kPa) (Fig. 4c).[13] When immersed overnight in water, the obtained gels swelled considerably and reached an equilibrium swelling ratio Q of ca. 110 (Fig. 4d), slightly increasing with precursor concentration between 2.5 and 10%.

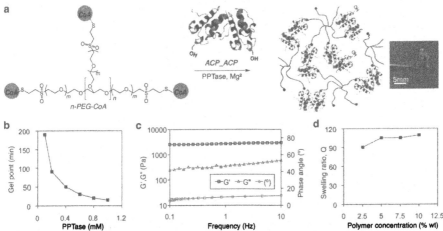

Figure 4.a, Scheme of PPTase-mediated cross-linking of PEG hydrogels and obtained 5% hydrogel (stained in blue) after swelling in water; **b,** time of hydrogel precursors gelation recorded as a function of enzyme concentration; **c,** oscillatory frequency sweep performed on cross-linked hydrogel shows G'>G", indicating the presence of a crosslinked elastic polymer network; **d,** the swelling ratio Q, determined as the swollen gel mass divided by the gel's dry mass. Figure reprinted with permission from Reference [14]. Copyright 2010 American Chemical Society.

Bio-functionalization of PPTase hydrogels

In order to expand the functionality of these PPTase-crosslinked semi-synthetic hydrogels, we explored the incorporation of a bioactive peptide as pendant moiety of the polymer network (Fig. 5a, b). The fibronectin-derived cell-adhesion peptide RGDS was chosen as cell adhesion ligand. RGDS was conjugated to CoA using the heterobifunctional PEG linker NHS-PEG-VS (Fig. 5a), rendering the resulting CoA-RGDS conjugate reactive towards

ACP_ACP during network formation. Notably, addition of 50 μM CoA-RGDS rendered PPTase hydrogels adhesive for primary fibroblasts whereas the negative control did not enable cell spreading (Fig. 5b, c).

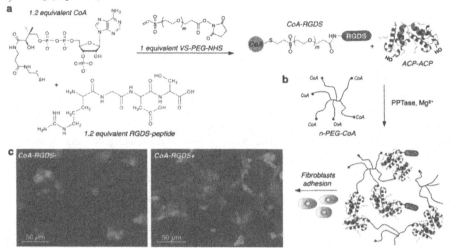

Figure 5: **a,** Scheme of CoA–RGDS conjugation using VS-PEG-NHS; **b,** CoA-RGDS conjugate is tethered to the forming network *via* PPTase; **c**: Green fluorescent protein (GFP)-expressing fibroblasts were culture for 2 hours on the hydrogels without CoA-RGDS (left) and within 50 μM of CoA-RGDS (right). Parts of the figure adapted with permission from Reference [14]. Copyright 2010 American Chemical Society.

3-D in vitro cell culture within PPTase hydrogels

Finally, we assessed the biocompatibility of the novel gels in 3D cell cultures using a live-dead assay. Primary mouse fibroblasts were encapsulated in gels by conducting the PPTase-mediated crosslinking *in situ* (Fig. 6). After 4 h of encapsulation cell viability was ca. 95%, not distinguishable from cultures in a well-established chemically crosslinked 3D gel system [13].

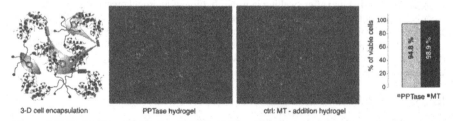

Figure 6. Cell viability of wild-type fibroblasts after 4 h of encapsulation determined by live/dead measurement. Fluorescent confocal images with standard green and red filters and 10x objective have approximately z of 400 μm depth. Imaris software was used for 3D reconstruction

and cell counting. PPTase hydrogels show high cell viability, comparable to gels crosslinked by Michael-type addition. Parts of the figure adapted with permission from Reference [14]. Copyright 2010 American Chemical Society.

CONCLUSIONS

We demonstrate that PPTase is suitable for selective hydrogel formation and its conjugation with bioactive moieties. The small size of the PPTase we used here, its high expression yields and ease of purification (using His-tag we obtained nearly 100% purity) as well as high specificity lets us conclude that the PPTase gels could be an attractive alternative to other enzymatically-crosslinked gel systems, such as those that are formed by the much larger and difficult-to-produce transglutaminases[4, 5, 8]. The resulting gels possess promising biological characteristics, which could be explored in a number of applications in cell biology and tissue engineering.

REFERENCES

[1] M. P. Lutolf, N. Tirelli, S. Cerritelli, L. Cavalli, J. A. Hubbell, *Bioconjug Chem* 12 (2001) 1051-1056.
[2] C. A. DeForest, B. D. Polizzotti, K. S. Anseth, *Nature Materials* 8 (2009) 659-664.
[3] B. H. Hu, J. Su, P. B. Messersmith, *Biomacromolecules* 10 (2009) 2194-2200.
[4] J. J. Sperinde, L. G. Griffith, *Macromolecules* 30 (1997) 5255-5264.
[5] M. Ehrbar, S. C. Rizzi, R. Schoenmakers, J. A. Hubbell, F. E. Weber, M. P. Lutolf, *Biomacromolecules* 8 (2007) 3000-3007.
[6] T. H. Chen, H. D. Embree, E. M. Brown, M. M. Taylor, G. F. Payne, *Biomaterials* 24 (2003) 2831-2841.
[7] M. Kurisawa, J. E. Chung, Y. Y. Yang, S. J. Gao, H. Uyama, *Chemical Communications* (2005) 4312-4314.
[8] B. H. Hu, P. B. Messersmith, *Journal of the American Chemical Society* 125 (2003) 14298-14299.
[9] R. H. Lambalot, A. M. Gehring, R. S. Flugel, P. Zuber, M. LaCelle, M. A. Marahiel, R. Reid, C. Khosla, C. T. Walsh, *Chemistry & Biology* 3 (1996) 923-936.
[10] Z. Zhou, P. Cironi, A. J. Lin, Y. Xu, S. Hrvatin, D. E. Golan, P. A. Silver, C. T. Walsh, J. Yin, *ACS Chem Biol* 2 (2007) 337-346.
[11] L. S. Wong, J. Thirlway, J. Micklefield, *Journal of the American Chemical Society* 130 (2008) 12456-12464.
[12] J. Yin, P. D. Straight, S. M. McLoughlin, Z. Zhou, A. J. Lin, D. E. Golan, N. L. Kelleher, R. Kolter, C. T. Walsh, *Proceedings of the National Academy of Sciences of the United States of America* 102 (2005) 15815-15820.
[13] M. P. Lutolf, J. A. Hubbell, *Biomacromolecules* 4 (2003) 713-722.
[14] K. A. Mosiewicz, K. Johnsson, M. P. Lutolf, *Journal of the American Chemical Society* DOI: 10.1021/ja9098164 (2010).
[15] A. Keppler, S. Gendreizig, T. Gronemeyer, H. Pick, H. Vogel, K. Johnsson, *Nature Biotechnology* 21 (2003) 86-89.

Mater. Res. Soc. Symp. Proc. Vol. 1272 © 2010 Materials Research Society   1272-PP07-05

Dynamic constitutional membranes-toward an adaptive facilitated transport

Mihail Barboiu*

Adaptive Supramolecular Nanosystems Group, Institut Européen des Membranes, ENSCM/ UMII/UMR-CNRS5635, Place Eugène Bataillon CC047, 34095 Montpellier Cedex 5, France. E-mail: mihai.barboiu@iemm.univ-montp2.fr

ABSTRACT

Dynamic Interactive Systems are defined by networks of continuously exchanging and reversibly reorganizing connected objects (supermolecules, polymers, biomolecules, pores, nanoplatforms, surfaces, liposomes, cells). They are operating under the natural selection to allow spatial / temporal and structural / functional adaptability in response to internal constitutionalor to stimulant external factors. In this minireview we will disscuss some selected examples of organic/inorganic SYSTEMS MATERIALS, covering a) the sol-gel resolution of constitutional architectures from Dynamic combinatorial libraries and b) the generation of Dynamic Hybrid Materials and SYSTEMS MEMBRANES able to evolve insidepore architectures via ionic stimuli so as to improve membrane transport functions.

INTRODUCTION

Constitutional Dynamic Chemistry (CDC) [1] and its application Dynamic Combinatorial Chemistry (DCC) [2] are new evolutional approaches to produce chemical diversity. In contrast to the stepwise methodology of classic combinatorial chemistry, DCC allows for the simple generation of diverse interexchanging architectures from sets of building molecules interacting reversibly. With the DCC approach, the building library elements are spontaneously assembled to form all possible virtual combinations. If libraries are produced in the presence of a specific target, new ligands can be selected that resemble the naturally occurring ones. In this way, new, potentially useful affinity ligands can be generated. Compound libraries-DCL generated by DCC show special interest on a very diverse range of applications: molecular and supramolecular recognition, drug-, catalyst- and material discovery. Kinetic or thermodynamic resolution self-assembly followed by covalent-modification or by crystallization shed light on useful strategies to control and to make convergence between self-organization and functions. [3] Basically, the CDC implements a reversible interface between interacting constituents, mediating the structural self-correlation of different domains of the system by virtue of their basic constitutional behaviours. The self-assembly of the components controlled by mastering supramolecular interactions, may allow *the flow of structural information from molecular level toward nanoscale dimensions.* Understanding and controlling such up-scale propagation of structural information, might offer the potential to impose further precise order at the mesoscale and new routes to obtain highly ordered ultradense arrays over macroscopic distances. Within this context, during the last decade the CDC is expressing more and more interest for *Dynamic Interactive Systems*, (DIS). Networks of continuously exchanging and reorganizing reversibly connected objects (supermolecules, polymers, biomolecules, pores, surfaces, nanoplatforms, liposomes, cells) form the core of DIS, operating under the natural selection to allow 4D spatial / temporal and structural / functional adaptability in response to internal constitutional or stimulant external factors. This makes DIS a great source of knowledge, highly relevant for a huge variety of direct applications. This minireview will discuss some selected examples of DIS concerning the hybrid organic/inorganic systems prepared by **sol-gel resolution of constitutional systems from DCLs** and then we will introduce the **Dynamic Hybrid Materials and Membranes** able to evolve architectures of functional behaviors.

Dynamic constitutional resolution by sol-gel-SYSYEMS MATERIALS

In contrast to the stepwise methodology of classic combinatorial techniques, dynamic combinatorial chemistry (DCC) allows for the simple generation of large chemical libraries from small sets of building blocks based on reversible interconversion between the library species. With this DCC approach, the building elements are spontaneously assembled to virtually encompass all possible combinations using dynamic covalent or non-covalent bonds between the species. A specific advantage with dynamically generated libraries addresses the possibility for the compounds to self-adjust to a chosen target at a given time in a certain environment. By virtue of the reversible interchanges, the library can adapt to the internal and external system constraints; for example allowing selection events driven by molecular recognition. In the past few years, different approaches to DCC have been reported for the resolution of dynamic molecular libraries (DCL), notably by freezing the reversible equilibria of DCL. Investigations on DCC have highlighted the wide potentialities of dynamic combinatorial resolution processes (DCR) of *molecular DCLs*, kinetically controlled in the second step by the irreversible enzymatic or tandem reactions. The control of *supramolecular DCLs* by a phase-change can be also controlled by using the sol-gel process, providing simple methods for synthesis of complex hybrid architectures. [4-6] The concept is presented in Fig. 1: a molecular library is allowed to form by reversible self-assembly different sets of dynamic supramolecular components in solution. Then a specific subset of the DCL is further amplified, due to its adaptation to the system constraints, for example allowing selection events driven by molecular recognition. The higher stability of one component may be the main driving force to form a solid hybrid network via the kinetically controlled sol-gel process. This natural selection allows structural/ functional adaptability of the hybrid system in response to internal constitutional (Fig.1a) or stimulant external factors (Fig.1b). Within this context, we have recently designed liquid, bilayer or solid hybrid membrane configurations, based on heteroditopic receptors. [4-7] They generate self-organized superstructures in solution and in the solid state based on three encoded features: 1) molecular recognition sites for the anion and the cation are covalently linked (Fig.2a), 2) supramolecular guiding interaction is the urea head-to-tail H-bond association, 3) the triethoxysilyl groups allow by sol-gel process to transcribe (to freeze) in a solid hybrid material the self-organized superstructures in solution. Such heteroditopic receptors generate organogels, revealing the formation of dynamic libraries of supramolecular oligomers in solution (Fig.2b). The hierarchical dynamic self-assembly of ribbon-type architectures lead in a second *sol-gel transcription/resolution* step to lamellar solid hybrid materials at nanoscopic scale (Fig.2c). Understanding and controlling such up-scale propagation of the supramolecular information might offer the potential to impose further precise order at the mesoscale. The resulted hybrid materials are composed from functional domains self-organized at the nanometric level, randomly ordered in the hybrid matrix. These oriented nanodomains result from the controlled self-assembly of simple molecular components which are then fixed by sol-gel transcription.

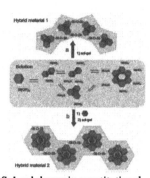

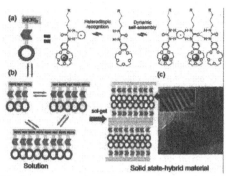

Fig. 1 Sol-gel dynamic constitutional resolution of the hybrid systems arrays under the pressure of the driving forces: (a) internal structural/ constitutional stabilization or (b) in response to a external target.

Fig.2 (a) Dynamic self-organization and salts recognition by ureido-macrocyclic receptors. (b) Generation of DSL of oligomers in solution and (c) sol-gel transcription/ resolution of solid state ion-conduction nano-ribbons (TEM image).

They can be successfully used to transfer the overall supramolecular functionality via their self-organization in functional transporting hybrid nanomaterials. For example, ion-exchanging IEM [4] or proton exchanging PEM hybrid membranes [5] presenting long-range lamellar (hundreds of nanometers) self-organization were recently surveyed for their ability to form functional-layers, reaching high ionic conductivities. These results provided new insights into the basic features that control the convergence of *supramolecular self-organization* with the conduction *supramolecular function* and suggest tools for developing the next generation of high conductive materials. This is reminiscent with the supramolecular organization of binding sites in channel-type proteins collectively contributing to the translocation of solutes along the hydrophilic ways.

Dynamic insidepore resolution towards Dynamic hybrid materials/Membranes

Unlike soft hybrid materials, mesoporous materials are often able to form oriented nanopores with well-defined and controled dimensions and geometries. Nanoscale confinement of organic/ inorganic components within integrated hybrid systems can induce collective specific properties with a different functionality , going beyond the properties of any of these components acting in solution or entraped in solid matrixes. Within this context, dynamic molecular/supramolecular libraries/networks can organize and more evolve in such scaffolding inorganic nanospace to develop new "dynamic system functions" related to confinement or anisotropic orientation. Thus as an outgrowth of Dynamic Constitutional Chemistry principles, it leads to a **dynamic intrapore resolution of DCLs**. These phenomena might be considered as an upregulation of the most adapted 3D insidepore/ super- or nanostructure under confined conditions, naturally adapting the system architecture/ functionality in response of various effectors.
It embodies a reorganization of the system spatial/temporal configuration evolving an improved response in the presence of the effectors that produced this change (Fig. 3). Operating such evolutive Dynamic Interactive Systems is mainly related to reversible interactions between the continually interchanging confined organic components and the inorganic porous preformed structure of specific directional behaviors. [11]

237

This makes them adaptive, responding to internal constitutional or to external stimuli. It would result in the formation of the most efficient functional superstructures, in the presence of the fittest stimulus, selected from a set of diverse less-selective possible architectures which can form by self-assembly, within such confined conditions. This can lead to an adaptive functionality, directed by the evolution of 'dynamic encodable' interactions between components *self-designing the available space under confinement.*

More generally, applying such consideration to hybrid constitutional systems, leads to the definition of *dynamic hybrid materials*, in which inorganic porous matrixes are reversibly connected with dynamic organic/supramolecular networks. They might provide new insights into the basic features that control the self-design of functional biomimetic systems.

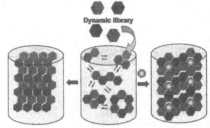

Fig. 3 Dynamic insidepore constitutional resolution of the systems arrays under the pressure of the driving forces :(a) internal structural/ constitutional stabilization or (b) in response to a external target.

SYSTEMS MEMBRANES

Many fundamental biological processes appear to depend on unique properties of inner hydrophylic domains of the membrane proteins in which ions or water molecules diffuse along the directional pathways. Bilayer or nanotube artificial membranes were developed with the hope to mimic the natural ion-channels, which can directly benefit the fields of separations, sensors or storage-delivery devices. Within this context constitutional self-instructed membranes were developed by our group and used for mimicking the adaptive structural functionality of natural ion-channel systems. [8] From the conceptual point of view and in a similar manner, columnar silica mesopores can be used as a scaffolding matrix to orient self-organized ion-channel-type artificial systems. These membranes are based on dynamic hybrid materials in which the functional self-organized macrocycles are reversibly connected with the inorganic silica through hydrophobic non-covalent interactions (Fig.4a). Supramolecular ion-channel architectures confined within scaffolding hydrophobic silica mesopores can be structurally determined and morphologically tuned by alkali salts templating. The dynamic character lied to reversible interactions between the continually interchanging components make them respond to external ionic stimuli and adjust to form the most efficient transporting superstructure, in the presence of the fittest cation, selected from a set of diverse less-selective possible architectures which can form by their self-assembly. Evidence has been presented that such membrane adapts and evolves its internal structure so as to improve its ion-transport properties: the dynamic non-covalent bonded macrocyclic ion-channel-type architectures can be morphologically tuned by alkali salts templating, during the transport experiments or the conditioning steps. From conceptual point of view these membranes express a synergistic adaptive behaviour: the addition of the fittest alkali ion drives a evolution of the membrane toward the selection and amplification of the specific transporting superstructures within the membrane in the presence of the cation that promoted its generation in the first place. This is a nice example of dynamic self-instructed ("trained") membranes where a solute induces the up-regulation (prepare itself) of its own selective membrane system.

238

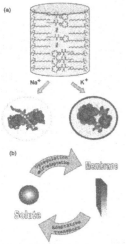

This led to the discovery of the functional supramolecular architecture evolving from a mixture of reversibly insidepore exchanging devices via ionic stimuli so as to improve membrane ion transport properties. These phenomena might be considered as an upregulation of the most adapted 3D *"insidepore"* superstructure, enhancing the membrane efficiency and the selectivity by the binding of the ion-effectors (Fig. 4b). Finally these results extend the application of *constitutional dynamic chemistry* from materials science to *functional Dynamic Interactive Membranes- SYSTEMS-MEMBRANES.*

Fig. 4 (a) Generation of directional ion-conduction pathways by hydrophobic confinement of ureido-macrocyclic receptors within silica mesopores. They are morphologically tunable by alkali salts templating within such *Dynamic hybrid materials.* (b) It embodies a reorganization of the membrane configuration evolving an improved response in the presence of the solute that produced this change in the first place.

This feature offers to membrane science perspectives towards self-designed materials evolving their own functional superstructure so as to improve their transport performances. Prospects for the future include the development of these original methodologies towards dynamic materials, presenting a greater degree of structural complexity. [9, 10] They might provide new insights into the basic features that control the design of materials mimicking the protein channels with applications in separations, sensors or as storage-delivery devices.

CONCLUSIONS

Highly interconnected networks relate to a Systems Chemistry. The self-assembly of the system components controlled by mastering molecular/ supramolecular interactions, may embodies the flow of structural information from molecular level toward nanoscale dimensions. Resultant spatio/temporal affinity is highly dependant on the nature of the scaffold as multivalent interactions have been shown to be influenced by such factors as shape, valency, orientation and flexibility. The possibility of designing and engineering nanometric multivalent platforms is at the forefront of cross-disciplinary oriented research. Thus one of the next challenges is to implement the "living" Dynamic Interactive Systems supporting natural selection and functional evolution as a viable solution to post-synthetically assembled systems. Natural systems have evolved for billion of years to accept complex evolutive structures, not synthetic generated systems. Of scientific significance is that Dynamic Interactive Systems could be extended to the vast field of scientific challenges, for example, resulting in the property (function)-driven generation of new adjustable (adaptive) system structures. Sol-gel selection of constitutional architectures from Dynamic combinatorial libraries and the generation of Dynamic Hybrid Materials and Membranes toward SYSTEMS MATERIALS and SYSTEMS MEMBRANES should expand the fundamental understanding of nanoscale structures and properties as it relates to creating products and manufacturing processes. Combined supramolecular and combinatorial dynamic strategies to produce nanosystems can be effectively shared as soon merged marketable nanotechnology to benefit as regards most of research laboratories and nanomaterials producers.

REFERENCES

1. J.-M. Lehn, *Chem. Soc. Rev.*, **36**, 151-160 (2007).
2. (a) J.-M. Lehn, *Chem. Eur. J.*, **5**, 2455-2463 (1999); (b) P. T. Corbett, J. Leclaire, L. Vial, K. R. West, J.-L. Wietor, J. K. M. Sanders, S. Otto, *Chem. Rev.* **106**, 3652-3711 (2006).
3. (a) F. Dumitru, E. Petit, A. van der Lee, M. Barboiu, *Eur. J. Inorg. Chem.*, 4255-4262 (2005); (b) Y. M. Legrand, A. van der Lee, M. Barboiu, *Inorg. Chem.*, **46**, 9540-9547 (2007) ; (c) M. Barboiu, E. Petit, A. van der Lee, G. Vaughan, *Inorg. Chem.*, **45**, 484-486 (2006) ; (d) F. Dumitru, Y.M. Legrand, A. van der Lee, M. Barboiu, *Chem. Commun*, 2667-2669 (2009); (e) M. Barboiu, F. Dumitru, Y.-M. Legrand, E. Petit, A. van der Lee, *Chem. Commun*, 2192-2194; (f) M. Barboiu, E. Petit, G. Vaughan, *Chem. Eur. J.*, 2004, **10**, 2263-2270 (2009); (g) C. Luca, M. Barboiu, C.T. Supuran J. M. Lehn, *Rev. Roum. Chim.*, **36**, 1169-1173 (1991); (h) M. Barboiu, J. M. Lehn, *Revista de Chimie*, **59**, 255-259 (2008) ; (i) M. Barboiu, J. M. Lehn, *Rev. Roum. Chim.*, **9**, 909-914 (2006); (j) G. Nasr, E. Petit, D. Vullo, J.Y. Winum, C.T. Supuran, M. Barboiu, *J. Med. Chem*, **42**, 4853-4859 (2009); (k) M. Barboiu, J.-M. Lehn, *Proc. Natl. Acad. Sci. USA*, **99**, 5201-5206 (2002); (l) M. Barboiu, G. Vaughan, R. Graff, J.-M. Lehn, *J. Am. Chem. Soc.*, **125**, 10257-10265 (2003); (m) M. Barboiu, E. Petit, G. Vaughan, *Chem. Eur. J.*, 2004, **10**, 2263-2270 (2004) ; (n) P. Vongvilai, M. Angelin, R. Larsson, O. Ramström, *Angew. Chem. Int. Ed.*, **46**, 948–950 (2007); (o) M. Angelin, P. Vongvilai, A. Fischer, O. Ramström, *J. Org. Chem.* **73**, 3593–3595 (2008).
4. M. Michau, R. Caraballo, C. Arnal-Hérault, M. Barboiu, *J. Membr. Sci.*, **321**, 22-30 (2008).
5. M. Michau, M. Barboiu, *J. Mater. Chem.* **19**, 6124-6131 (2009).
6. a) S. Mihai, A. Cazacu, C.Arnal-Herault, G. Nasr, A. Meffre, A. van der Lee, M. Barboiu, *New. J. Chem.*, **33**, 2335-2343 (2009); b) E. Mahon, T. Aastrup, M. Barboiu, *Chem. Commun.* 2010, DOI; 10.1039/b924766a; c) J. Nasr, M. Barboiu, T. Ono, S. Fujii, J.-M. Lehn, *J. Membr. Sci.*, **321**, 8-14 (2008).
7. a) M. Barboiu, G. Vaughan, A. van der Lee, *Org. Lett.* 2003, **5**, 3073-3076; (b) M. Barboiu, *J. Incl. Phenom. Mol. Rec.* **49**, 133-137 (2004); (c) A. Cazacu, C. Tong, A. van der Lee, T. M. Fyles, M. Barboiu, *J. Am. Chem. Soc.* **128**, 9541-9548 (2006); (d) M. Barboiu, S. Cerneaux, A. van der Lee, G. Vaughan, *J. Am. Chem. Soc.* **126**, 3545-3550 (2004); e) C. Arnal-Herault, M. Michau, A. Pasc-Banu, M. Barboiu, *Angew. Chem. Int. Ed.*, **46**, 4268-4272 (2007); f) C. Arnal-Herault, A. Pasc-Banu, M. Michau, D. Cot, E. Petit, M. Barboiu, *Angew. Chem. Int. Ed.* **46**, 8409-8413 (2007); g) C. Arnal-Hérault, M. Barboiu, A. Pasc, M. Michau, P. Perriat, A. van der Lee, *Chem. Eur. J.*,**13**, 6792-6800 (2007).
8. A. Cazacu, Y.M. Legrand, A. Pasc, G. Nasr, A. van der Lee, E. Mahon, M. Barboiu, *Proc. Natl. Acad. Sci.*, 2009, **106**, 8117-81.
9. E. Prouzet, S. Ravaine, C. Sanchez, R. Backov, *New J. Chem.*, **32**, 1284 – 1299 (2008); (b) S. Mann, *Nature Mat.* **8**, 781-792 (2009); (c) S. Mann, *Angew. Chem. Int. Ed.*, **47**, 5306-5320 (2008); (d) P. A. Wender, B.L. Miller, *Nature*, , **460**, 197-201 (2009).
10. (a) S. Xu, N. Giuseppone, *J. Am. Chem. Soc.*, **130**, 1826-1827 (2008); (b) R. Nguyen, L. Allouche, E. Buhler, N. Giuseppone, *Angew. Chem. Int. Ed.*, **48**, 1093-1096 (2009); (c) R. Nguyen, E. Buhler, N. Giuseppone, *Macromolecules*, **42**, 5913-5915 (2009); (d) L. Tauk, A. P. Schröder, G. Decher, N. Giuseppone, *Nature Chem.*, **1**, 649-6542 (2009).
11. A. Guerrero-Martínez, S. Fibikar, I. Pastoriza-Santos, L. M. Liz-Marzán, L. De Cola, *Angew. Chem. Int. Ed.* **48**, 1266-1270 (2005); (b) M. Rosso-Vasic, L. De Cola, H. Zuilhof, *J. Phys. Chem. C* **113**, 2235–2240 (2009).

AUTHOR INDEX

SUBJECT INDEX

Printed in the United States
by Baker & Taylor Publisher Services